'Dr Sanjay Gupta has the unique a [...] research into accessible language w [...] perspective, and practical strategies [...] comprehensive guide for people of any age concerned about their memory and cognitive health.' Gary Small, MD, Director, UCLA Longevity Center, and author of *The Small Guide to Alzheimer's Disease*

'A must-read owner's manual for anyone with a brain.' Arianna Huffington

'Dr Gupta is brilliant at busting myths, allaying fears, and giving us the solutions we need to keep sharp throughout our lives.' Dr Dean Ornish, founder and president, Preventive Medicine Research Institute

'I wish *Keep Sharp* existed when I was helping take care of my father as he suffered from Alzheimer's.' Maria Shriver, founder, The Women's Alzheimer's Movement

'*Keep Sharp* will change how you think about your brain, how you care for it, and how you support its full potential.' Dr David B. Agus, *New York Times* bestselling author of *The End of Illness*

'Dr Sanjay Gupta brilliantly crafts an actionable plan we can all follow. I feel sharper already.' Dr Mehmet Oz, host, The Dr. Oz Show

'This is the book all of us need, young and old!' Walter Isaacson, *New York Times* bestselling author of *Steve Jobs*

'*Keep Sharp* looks at the lessons we can learn from so-called 'super brained' people in their 80s and 90s while debunking some common myths about ageing and cognitive decline.' *Eastern Daily Press*

'Offers many surprising suggestions. . . not only will this book give you a better idea of how your brain works, it will help it to work better too.' *Daily Express*

Dr Sanjay Gupta

KEEP SHARP

AARP®

Real Possibilities

Copyright © 2021 Dr Sanjay Gupta

AARP and Staying Sharp are registered trademarks of AARP. All rights reserved.

The right of Dr Sanjay Gupta to be identified as the Author of the Work has been asserted by him in accordance with the Copyright, Designs and Patents Act 1988.

First published in 2021 by SIMON & SCHUSTER US

First published in Great Britain in paperback in 2022 by HEADLINE HOME an imprint of HEADLINE PUBLISHING GROUP

3

Apart from any use permitted under UK copyright law, this publication may only be reproduced, stored, or transmitted, in any form, or by any means, with prior permission in writing of the publishers or, in the case of reprographic production, in accordance with the terms of licences issued by the Copyright Licensing Agency.

Every effort has been made to fulfil requirements with regard to reproducing copyright material. The author and publisher will be glad to rectify any omissions at the earliest opportunity.

Cataloguing in Publication Data is available from the British Library

ISBN 978 1 4722 7423 6
e-ISBN 978 1 4722 7422 9

Interior design by Carly Loman

Printed and bound in India by Manipal Technologies Limited, Manipal

MIX
Paper from responsible sources
FSC™ C104740

HEADLINE PUBLISHING GROUP
An Hachette UK Company
Carmelite House
50 Victoria Embankment
London
EC4Y 0DZ

www.headline.co.uk
www.hachette.co.uk

For my three girls, Sage, Sky, and Soleil. In order of age, so as to preempt any future disputes over the dedication order. I love you so much, and watched you grow faster than this book. Always take the time to be completely present, because it is perhaps the best and most joyous way to keep your mind sharp and your life bright. You are still so young, yet you have given me a lifetime of memories I hope to never forget.

For my Rebecca, who has never wavered in enthusiasm. If in the end, our lives are just a collection of memories, mine will be filled of images of your beautiful smile and your steadfast support.

For anybody who has dreamed that their brain can be better. Not just free of disease or trauma, but optimized in a way that allows you to best build and remember your life narrative, and equips you to be resilient through life's challenges. For anyone who has always believed their brain wasn't a black box, impenetrable and untouchable, but could be nourished and grown into something greater than they imagined.

Remembrance of things past is not necessarily
the remembrance of things as they were.

– Marcel Proust

Contents

KEEP SHARP

Nothing Brainy about It

The brain is wider than the sky ... [and] ... deeper than the sea.

EMILY DICKINSON

Unlike most of my colleagues, I didn't grow up with a deep-seated desire to be a doctor, let alone a brain surgeon. My earliest aspiration was to be a writer, likely triggered by a boyhood crush I had on a grade school English teacher. When I chose medicine, I was thirteen years old and my grandfather had just suffered a stroke. We were very close, and witnessing his brain function change so quickly was jarring. He was suddenly unable to speak or write but seemed to understand what people said and could read without difficulty. Simply put, he could receive verbal and written communication easily, but he could not respond in those same ways. It was the first time I became fascinated by the intricate and mysterious functioning of the brain. I spent a lot of time at the hospital and was that annoying kid who asked the doctors a lot of questions. I felt very grown up as they patiently explained what had happened. I watched as those doctors were able to return my grandfather to good health after opening up his carotid artery to restore the blood flow to his brain and prevent future strokes. Having never spent much time with surgeons before then, I was hooked. I started reading everything I could about medicine and the human body. Before long, I was fixated on the brain, and specifically memory. It still astonishes me that our memories—the very fabric of who we are—can be reduced to invisible neurochemical signals between tiny areas of the brain. For me, those early explorations into the world of brain biology were at once demystifying and magical.

Years later, when I was in medical school in the early 1990s, conventional wisdom was that brain cells, such as neurons, were incapable of regenerating. We were born with a fixed set and that was it; throughout life, we'd slowly drain the cache (and accelerate that killing off with bad habits like drinking too much alcohol and smoking marijuana—the truth about that later). Perhaps it was the eternal optimist in me, but I never believed that our brain cells simply stopped growing and regenerating. After all, we continue to have novel thoughts, deep experiences, vivid memories, and new learning throughout our lives. It seemed to me that the brain wouldn't just wither away unless it was no longer being used. By the time I finished my neurosurgery training in 2000, there was plenty of evidence that we could nurture the birth of new brain cells (called neurogenesis) and even increase the size of our brains. It was a staggeringly optimistic change in how we view the master control system of our bodies. Indeed, every day of your life, you can make your brain better, faster, fitter, and, yes, *sharper*. I am convinced of that. (I'll get to the bad habits later; they don't necessarily kill brain cells, but when they are abused, they can alter the brain, especially its memory powers.)

Let me say at the outset: I am certainly a fan of excellent education, but this is not what *Keep Sharp* is all about. This book is less about improving intelligence or IQ and more about both propagating new brain cells and making the ones you have work more efficiently. This isn't so much about remembering a list of items, performing well on exams, or executing tasks adeptly (though all of those goals will be more achievable with a better brain). In *Keep Sharp*, you will learn to build a brain that connects patterns others might miss and helps you better navigate life. You will develop a brain able to toggle back and forth between short-term and long-term views of the world and, perhaps most important, a brain highly resilient in the face of life experiences that might be disabling to someone else. In this book, I will precisely define *resilience* and teach you how to nurture it. Resilience has been a critical ingredient for my own personal growth.

Context matters when talking about something as important as the function or dysfunction of our brains, and our view of cognitive decline has changed dramatically over time. The history of documenting dementia dates back to at least 1550 BCE, when Egyptian doctors first described the disorder in what's known as the Ebers Papyrus, a 110-page scroll or manuscript that contains a record of ancient Egyptian medicine. But it was not until 1797 that the phenomenon was given a name, *dementia*, which literally means "out of one's mind" in Latin. The term was coined by a French psychiatrist, Philippe Pinel, who is revered as the father of modern psychiatry for his efforts to push for a more humane approach to the care of psychiatric patients. When the word was first used, however, *dementia* referred to people with an intellectual deficit ("abolition of thinking") at any age. It was not until the end of the nineteenth century that the word was confined to people with a specific loss of cognitive ability. During that century, the British physician Dr. James Cowles Prichard also introduced the term *senile dementia* in his book, *A Treatise on Insanity*. The word *senile*, which means old, referred to any type of insanity that occurs in old people. Because memory loss is one of the most prominent symptoms of dementia, the word became mostly associated with old age.

For a long time, the elderly with dementia were believed to be cursed, or to have an infection like syphilis (because the symptoms of syphilis can be similar). So the word *dementia* was considered pejorative, used as an insult. In fact, when I first told my kids I was writing this book, they asked if it was about *dementors*, the dark, soul-sucking creatures from Harry Potter. The idea that dementia, which is not a specific disease but a group of symptoms associated with memory loss and poor judgment, is sometimes thought of in such negative ways is worth addressing briefly here.

It is true that scientists and doctors use the word clinically, and it is also true that patients and their loved ones don't always know what to make of it, especially when they first receive the diagnosis. It is too imprecise, for one thing. Dementia can be a spectrum, ranging from mild to severe, and some of the causes of dementia are entirely reversible. Alzheimer's disease, which accounts for more than half the cases of dementia, gets nearly all the attention, and as a result, the terms *dementia* and *Alzheimer's* are often used interchangeably. They shouldn't be. The word *dementia*, however, is steeped in our common vernacular, and so is the association with Alzheimer's disease. In this book, I use both terms with the hope that the conversation, and the words we use to describe the broad condition of cognitive decline, will shift in the future.

I believe there has been an overemphasis on Alzheimer's disease as a way to talk about this broad condition, and it has further fueled a widespread sense of fear that memory loss is inevitable as we get older. Perfectly healthy people in their thirties and forties are alarmed about the implications of common memory lapses, like misplacing their keys or forgetting someone's name. That is a misguided fear, and as you will learn, memory loss is not a preordained part of aging.

As I started traveling the world talking to people about this book, I realized something else extraordinary. According to an AARP survey of Americans aged thirty-four to seventy-five, nearly everyone (93 percent) understands the vital importance of brain health, but those same people typically have no idea how to make their brains healthier or that achieving such a goal is even possible. Most believe this mysterious organ encased in bone is a black box of sorts, untouchable and incapable of being improved. Not true. The brain can be continuously and consistently enriched throughout your life no matter your age or access to resources. I have opened the black box and touched the human brain, and I will tell you all about those extraordinary experiences in this book. As a result of this training and decades of additional learning, I am more convinced than ever that the brain can be constructively changed—enhanced and fine-tuned. Just consider that. You probably think of your muscles that way—even your heart, which is a muscle. If you are reading this book, you are someone who is probably already proactive about your physical health. It is time to realize the same is possible with your brain. You can affect your brain's thinking and memory far more than you realize or appreciate, and the vast majority of people haven't even begun to try. *Keep Sharp* is going to help you design your own "sharp brain" program, which you can easily incorporate into your daily life. I have already done it myself, and I am excited to teach you to do it as well.

As an academic neurosurgeon and a reporter, a big part of my job is to educate and explain. I have learned that in order for my messages to stick, explaining the why of something is just as important as the what or the how. So throughout this book, I explain *why* your brain works the way it does and *why* it sometimes fails to deliver what you'd hoped. Once you understand these inner workings, the specific habits I encourage you to adopt will make sense and more likely become an effortless part of your routine.

Truth is, even when it comes to our general physical health, there is very little explanation in public discourse of how our bodies actually work and what makes them work better. Even worse, there is a lack of agreement among medical professionals about the best foods to eat, the types of activities we should pursue, or the amount of sleep we really need. It is part of the reason there are so many conflicting messages out there. Coffee is practically a superfood one day, and the next it's a potential carcinogen. Gluten is hotly debated continuously. Curcumin, found in turmeric, is touted as a miracle brain food, but what does that really mean? Statins seem to have a split personality, at least in the research circles: Some studies propose that statins lower risk for dementia and improve cognitive function, and other studies suggest the exact opposite. Vitamin D supplementation is constantly under fire too; some people swear by it, but study after study shows no benefit.

How does the average person make sense of the competing messages? Almost everyone agrees that toxins and pathogens from mercury to mold are bad for you, but what about certain artificial ingredients or even your own tap water? A new Canadian study showed that the fluoride in tap water consumed by pregnant mothers can lead to a small drop in their children's IQ later in life.[1] But fluorinated water also clearly has benefits for oral health and is still recommended by most top medical associations. It can be confusing. On top of all that, just about every doctor's visit ends with the blanket, generic recommendation that you should "get plenty of rest, eat right, and exercise." Sound familiar? Sure, it's good advice, but the problem is that there is hardly any consensus on what that means from a highly practical, day-by-day standpoint. What is the ideal diet, and how does it change from person to person? How about activity? High intensity, or slow and steady? Does everyone really need seven to eight hours of sleep a night, or can some people do just fine with far less? Why? Which drugs and supplements should one consider, given individual risk factors? And with brain health in particular, there is an even greater lack of basic understanding by both patients and the medical community. Has a doctor ever told you to take good care of your brain besides reminding you of the importance of wearing a helmet when riding a bike? Probably not.

Well, this doctor is going to tell you what you need to know and show you how to do it. If you think this already sounds complicated, don't worry. I am going to take you through step-by-step. You will understand more about your brain than you ever have in the past, and the ways to keep it healthy will make complete sense by the time you finish this book. Think of this as a master class on how to build a better brain, which opens the door to whatever you want to get out of life—including being a better father, mother, daughter, or son. You can be more productive and joyful, as well as more present for everyone with whom you interact. You will also develop more of that critical ingredient, resilience, so the optimization of your brain isn't derailed by the trials of daily life. These goals are all far more connected than you may realize.

Believing you can always be better tomorrow is an audacious way to view the world, but one that has helped shape my own life. Since I was a teenager, I've always worked hard on my physical health—to make my body stronger, faster, and more resilient to illness and injury. I think everyone has different motivations for taking care of their own health. For many, it is to feel better and more productive, and to be there for the children. For others, it is about achieving a certain physical appearance.

As we get older, the inspiration often comes from a brush with mortality and seeing the fragility of life up close. That was the case for me. When my father was just forty-seven years old, he developed crushing chest pain while out on a walk. I remember the panicked call I received from my mom, and the voice of the 911 operator I spoke to seconds later. A few hours later, he had an emergency four-vessel bypass operation on his heart. It was a frightening ordeal for our family, and we were worried he might not survive the operation. I was a young medical student at the time and fairly convinced I had somehow failed him. After all, I should've seen the warning signs, counseled him on his health, and helped him avoid heart disease. Luckily he survived, and the near-miss completely changed his life. He lost thirty pounds, paid close attention to the foods he was eating, and made regular activity a priority.

Now that I am past that age with my own children, I make it a priority to learn not just how to prevent disease, but to continually assess myself to make sure I'm performing to the best of my ability. Over the past few decades, I have also been exploring the deep connection between the heart and the brain. It is true that what is good for one is also good for the other, but I now believe the secret is that it all begins with your brain. As you are about to learn, once your brain is running cleanly and smoothly, everything else follows. You will make better decisions, have improved resilience and a more optimistic attitude, and the physical part of your body will improve too. There are studies showing that your pain tolerance will increase, your need for medications will decrease, and your ability to heal will be accelerated. Nearly every doctor I have spoken to about this book has said some variation of the following: In order to best take care of your body, you have to first take care of your mind. It is true, and the best part is that it is not that hard to do. Think of it as periodic little tweaks and adjustments instead of wholesale changes in your life.

Before I explain what those tweaks are and why they work, I offer my philosophy when it comes to the tone of this book. Over the years, I have worked in many different areas of our society: academic neuro-surgery jobs within universities; public service at the White House; as a journalist in media organizations; as a husband and father of three strong, smart, and beautiful girls. All along I have stuck to a principle I learned at a young age: Don't try to inspire people with fear. It doesn't work well, and it doesn't last long. When you scare someone, you activate that person's amygdala, the emotional center of the brain. The reaction is swift and hot, as one would have when confronted with a threat. The problem is that an action that starts in the emotional centers of the brain bypasses the judgment and executive function areas of the brain as well. As a result, the reaction may be intense and immediate, but it is also often uncoordinated and transient. It is why telling people they will likely have a heart attack if they don't lose weight may lead to a single intense week of dieting and exercise followed by an abrupt return to the old bad habits. Fear-based messaging will never lead to a long-term effective strategy because it is not the way we are wired. Nowhere is this more important than when telling someone he or she might develop Alzheimer's disease.

Many polls consistently show that people fear losing their minds more than anything else, even death. For many, it is considered the bogeyman of old age. And at one point in my life, I also worried a great deal about cognitive decline and dementia as I watched another grandparent progress through the stages of Alzheimer's disease. At first, he seemed to be contributing to conversations in nonsensical ways. Because he was a fun-loving, quick-to-laugh sort of guy, we thought perhaps he was making jokes we weren't quite in on yet. What finally gave him away was the vacant stare that would turn to puzzlement, and then to panic, as he realized he could not recall how to perform the most basic tasks and execute plans. I will never forget that look—at least I hope to never forget it.

But again, the fear of dementia should not be the motivation for you to read this book. Instead, it should be the knowledge that you can build a better brain at any age. I will teach you how to do it and explain why these strategies work. As you read this book, I don't want you to be running away from something. I want you to be running *toward* something— running toward a brain in peak shape that can withstand the test of your time on this planet.

When I started my work as a neurosurgeon more than twenty-five years ago, the idea of "improving" my own brain seemed like a bit of a misguided quest. After all, I was trained to remove tumors, clip aneurysms, relieve pressure from collections of blood and fluid, and so on. Even today, it is not possible for any neurosurgeon to go inside a human brain and adjust the 100 billion or so neurons to make the organ more intelligent and less vulnerable to decline. While heart surgeons can rotor away plaques in the heart, I can't rotor away the brain tangles often associated with Alzheimer's disease. There is no operation or medication to cure dementia or render someone more brilliant, creative, equipped with an extraordinary memory, or poised to invent the next big thing the world needs.

The brain is unlike any other organ. You can't transplant a brain like you can a heart (or liver, or kidney, or face for that matter), and our knowledge of the brain is still in the early stages, continuing to develop and expand. I had an astonishing realization recently while moderating a panel for the American Academy of Neurological Surgery, with the world's experts on concussion. They came from medicine, the Department of Defense, and the tech world. While they all talked about the great strides we have made in awareness, astonishingly there was no clear agreement on how to best treat a concussion, a condition diagnosed millions of times every year in the United States. There was hardly any published data presented to the Academy on effective treatments, either. Many current recommendations are based only on anecdotal evidence.[2] Even topics like rest—how much and for how long you rest a brain touched by concussion—were debated. For example, do you minimize activities that require concentration and attention during recovery from a concussion or increase those activities? When does light exercise, such as speed walking on a treadmill, help rather than hinder the recovery process? I heard all kinds of opinions, but very little of it was evidence based. And remember: That panel was made up of the world experts on brain injury.

Sure, we've come a long way from the ancient days of Aristotle, who thought the heart was the seat of intelligence and the brain was a kind of refrigerator that cooled off the fiery heart and hot blood, but there are still more questions about the brain than answers. We now know how actions are created and how thoughts are formed, and we can even identify the hippocampus, the two tiny seahorse-shaped structures in the brain essential for memory function. But we still haven't made much progress in stemming the tide of people with cognitive decline and dementia. While we enjoy lower rates of cardiovascular disease and certain cancers than a generation ago, the numbers are going in the other direction when it comes to brain-related impairment. According to a 2017 study out of the University of California, Los Angeles, 47 million Americans have some evidence of preclinical Alzheimer's disease, which means their brains show signs of adverse changes but symptoms have not yet developed. Oftentimes, it could still take years before their memory, thinking, and behavior are obviously affected.[3] The problem is we don't necessarily know who those 47 million people are and which ones will go on to develop full-blown Alzheimer's disease. We do know, however, that by 2060, the number of Americans with Alzheimer's dementia or cognitive impairment is expected to climb from 6 million to 15 million.[4] One new case of dementia will be diagnosed every four seconds, and it will be the most common neurodegenerative disorder of our time. Globally, the numbers of people living with Alzheimer's disease will swell to 152 million by 2050, which reflects a 200 percent increase since 2018. While science is steadily trying to push back, there still hasn't been a single new treatment since 2002 despite more than four hundred clinical trials for the disease.[5] That's why the gap between brain science and good therapeutics in drug discovery for brain disorders has been called the "valley of death."[6] That's the bad news.

The good news is that even without some major medical breakthrough, we can significantly optimize our brains in a variety of ways to improve its functionality, boost its neuronal networks, stimulate the growth of new neurons, and help stave off age-related brain illnesses. As you read this book, always remember this: Cognitive decline is not inevitable. As an analogy, think of a historical building that still stands. Perhaps it's more than a century old. Had it not been cared for throughout the decades, the wear and tear of weather and constant use would have certainly caused its deterioration and dilapidation. But with routine maintenance and occasional renovations, it not only withstood the test of time but is likely celebrated for its beauty, significance, and prominence. The same holds true with your brain, which is just another structure of different components and input needs for general maintenance and upkeep. Some of the strategies I will teach you will help assemble brain scaffolding—creating a support structure for your brain that is stronger and more stable than what you currently have and will help you to perform some initial "renovations," including reinforcement of your brain's "foundation." Other strategies will act to provide the raw materials necessary to do ongoing maintenance, as well as build what's called "cognitive reserve," or what scientists call "brain resiliency." With more cognitive reserve, you can lower your risk of developing dementia. Finally, there will be strategies that serve as finishing daily touches akin to dusting and tidying up to keep the better brain—better. As I mentioned, the old-school thinking dictated that the brain was pretty much fixed and hardwired after childhood development. Today, as we visualize the brain with new imaging technologies and study its ever changing function, we know the truth.

When you think of your heart, you probably have a good idea of things that might damage it: certain types of food, lack of activity, elevated cholesterol. But what about your brain? While many of those same things apply, your brain is also a highly sensitive antenna taking in millions of stimuli every day, and how we process these inputs can make a world of difference when it comes to a sharper brain. For example, I know many people who are absolutely crushed by events in the news, while others are emboldened and undaunted. Your brain can be strengthened by what you experience, like a good workout, or it can be battered and defeated. What separates those two camps of people? The answer is resilience. A resilient brain can withstand ongoing trauma, think differently, stave off brain-related illnesses including depression, and retain cognitive memory for peak performance.

Moreover, possessing a resilient brain is what separates strategic, visionary thinkers from more average ones. It is not necessarily IQ or even educational level. It is the ability to improve the brain from challenging experiences instead of shrinking it. Now *those* capabilities alone should be enough to motivate you to build a better brain. If you're looking to achieve your greatest potential, this book is for you. If you're hoping to gain insights into preventing the cognitive decline or dementia that affected someone in your family, this book is for you. (We know now that diseases like Alzheimer's start twenty to thirty years before any symptoms develop, so young people need to heed these lessons.) And if you're just seeking strategies to maximize brain health so that you can enjoy life to the fullest and be "unreasonably productive" no matter how young you are, this book also is for you. Whether you are dealing with a chronic disease or are an elite athlete, tomorrow can be better. Truth is, most of us—myself included—have not done nearly enough to improve ourselves. While writing this book, I tried everything I am recommending to you, and my brain has never been sharper. I want the same for you, and I will convince you that even small, incremental tweaks can have huge payoffs.

In 2017, I began collaborating with AARP (which changed its name from the "American Association of Retired Persons" because it speaks to a broader audience now and a lot of people never "retire"). Like me, AARP detects a sense of urgency around this book. They know that people are afraid of aging brains and of losing not only their cognition but also their freedom to live independently. AARP has established the Global Council on Brain Health to bring together scientists, health professionals, scholars, and policy experts from around the world. The goal is to gather the best possible advice about what we can be doing to maintain and improve brain health. The Council is chaired by Dr. Marilyn Albert, professor of neurology at Johns Hopkins University School of Medicine and director of the Division of Cognitive Neuroscience.

Since 2016, the Global Council has brought together ninety-four experts from twenty-three different countries and eighty different universities and organizations to reach a consensus on the state of the science. Together with fifty liaisons from government and not-for-profit organizations, the Global Council has produced a library of reports distilling evidence on how lifestyle and modifiable risk factors can impact brain health. So as part of our collaboration, I decided to put all of that wisdom—and a lot more—in these pages. I also spoke to people who have been directly affected by dementia and others who have spent their lives trying to understand and treat it. Through it all, I used my own lifelong fascination with and understanding of the brain to distill the enormous amount of information out there into one book with the insights and strategies you need to keep your mind sharp. Some of it will surprise you. I will be debunking a lot of myths that you've probably bought into and show you exactly what you could be doing this very minute to think and be sharper tomorrow. (Sneak peek: Stop multitasking. Don't spend mornings reading emails. Socialize more. Pick up the one specific activity that has long been scientifically proven to directly improve brain health—see chapter 4.) If I suggest something that's controversial (there can be lots of rival ideas in the brain health sphere), I will tell you. Trouble is, when science lacks universally accepted proof from long-term data, what can take hold—for good or bad—are theories, opinions, and perspectives.

You are going to hear this word a lot in this book: *lifestyle*. If there's one fact that's increasingly becoming apparent in scientific circles, it's that we are not doomed by the genetic cards we were dealt at birth. If a certain disease runs in your family, you can still stack the deck in your favor and avoid that fate. Our everyday experiences, including what we eat, how much we exercise, with whom we socialize, what challenges we face, how well we sleep, and what we do to reduce stress and learn, factor much more into our brain health and overall wellness than we can imagine. Here is an illustrative and fascinating example. A new study in 2018, published in the journal *Genetics*, revealed that the person we marry factors greater into our longevity than our genetic inheritance does.[6] And by a long shot! Why? Because it turns out that our lifestyle habits weigh heavily into our decisions around marriage—much more so than most other decisions in our lives. The researchers, who also analyzed birth and death dates of nearly 55 million family trees that encompassed 406 million people who were born from the nineteenth century to the mid-twentieth century, found that genes accounted for well under 7 percent of people's life span versus the 20 to 30 percent of most previous estimates. That means that over 90 percent of our health and longevity is in our own hands.

When I gathered all the highlights from my research colleagues at the Alzheimer's Association International Conference in 2019, one fact stood out: Clean living can slash your risk of developing a serious mind-destroying disorder, including Alzheimer's disease, even if you carry genetic risk factors. No matter what your DNA says, a good diet, regular exercise, not smoking, limiting alcohol, and some other surprising lifestyle decisions, can change that destiny. A few years ago, I experienced firsthand that healthy living could help someone overcome genetic risk for heart disease. Now we know the same is true for dementia. So worry less about your genes and stop using them as an excuse. Instead, focus on the things you get to choose, big and small, day in and day out.

I believe the way we have long approached the care of our bodies and brains is too passive. For much of medical history, doctors did nothing more than wait for disease or dysfunction to occur, and then they swooped in with antidotes to symptoms but not to the underlying pathology. As we evolved and cultivated more knowledge, we discovered that we could detect and diagnose disease before it reached late stages. Still, hardly anything was done to anticipate the disease long before it surfaced. Over the past few decades, we have begun to focus more attention on early intervention of disease and, more recently, on prevention. But in the brain health realm, attention to these final two areas has still been weak and too often missing. Let's change that. I wholeheartedly believe—and I am not alone in this thinking—that addressing brain decline is going to come from those two camps: prevention and early intervention. And I will throw another into the mix: optimization, or continuously building a better and more resilient brain.

Numerous books have been written about enhancing brain function and long-term brain health, but many of them are biased toward a particular philosophy, lack real data behind them, and are limited in their advice. The ones that I find particularly concerning are brain books that are platforms for selling products. The only thing I am selling (besides this book) is a way to understand your own brain and make it better. My aim is to present a comprehensive review of the science with practical lessons anyone can carry out starting now. I am not tied to any single "do this, not that" approach, though I do offer a few hard-and-fast rules. Like you, I seek the best that science has to offer, but the guidance has to also be truly practical.

I want you to keep one caveat in mind as you read this book: What will help stave off your cognitive decline may not be the same for someone else. If there's one fact I've learned in my years of studying the brain, operating on brains, and working with top scientists, it's that each of us carries our own unique profile. That is why any program to optimize brain health needs to be wide ranging, inclusive, and based on indisputable evidence. That is what I deliver in this book. And while there is no single nostrum, no one-size-fits-all solution (don't believe anyone who tells you otherwise), there are simple interventions all of us can make right away that can have a significant impact on our cognitive function and long-term brain health.

I am excited to share all the latest research and give you a personalized road map for arriving at a sharper brain for life. It is a spectacular destination.

IN THIS BOOK

For most of us, our brains are probably working at 50 percent capacity at any given time. That is my own made-up number. I don't know exactly how much it is (and neither does anyone else), but it is clear that with various behavior interventions such as meditation training or regular sound sleep, our brains can be put into hyperdrive status (and no, we of course don't use just 10 percent of our brains—see chapter 3). We know our brain can crank out a lot more torque than it is normally doing. So are our brains more like the mother whose child is pinned under a car and she displays superhuman strength to save him? Or are our brains high-performance Ferraris sputtering along on potholed neighborhood roads and hardly ever opening up to full throttle? I think it is the latter. We don't hit the open road enough with our exquisitely designed brains, and after a while, we forget what our brains are really capable of achieving.

You will read some car references in this book because they reflect the way I was raised. Both of my parents were in the auto industry; my mom was the first woman ever hired as an engineer by Ford Motor Company. So on my childhood weekends, often the whole family was tinkering with the family car. Our garage was filled with toolboxes and a running commentary about how the human body wasn't really that much differ-ent from the Ford LTD we were rebuilding. Both had engines, pumps, and life-sustaining fuel. I think those conversations contributed to my interest in the brain, because here was one area of the body that really couldn't be compared mechanically to a car. After all, there is no seat of consciousness in a car, no matter how plush the leather. Still, it is nearly impossible for me to look at the brain and not think about tuning up and maintenance. Is an oil change necessary? Is it getting the right fuel? Is it revving too high or being driven without a break? Are there cracks in the windshield or the chassis, and do all the tires have enough air pressure? Can it heat and cool properly? Does the engine respond appropriately to a sudden demand for speed, and how quickly can it be brought to a halt?

Part 1 starts with some basic facts. What exactly is a brain? What is it like to operate on one? What does it really look and feel like? Why is it so mysterious and hard to understand? How does memory work? What is the difference between normal brain aging and the occasional brain lapse, abnormal brain aging, and signs of serious decline? Then we'll take a deep dive into the myths about aging and cognitive decline, as well as how we know the brain can remodel, rewire, and grow.

Part 2 offers a tour of the five main categories that encompass all the practical strategies you need to protect and heighten your brain function: 1) exercise and movement; 2) sense of purpose, learning, and discovery; 3) sleep and relaxation; 4) nutrition; and 5) social connection. This part includes a look at some of the research going on now to explore the brain and find ways to better maintain and treat it. You'll meet top scientists who have dedicated their lives to decoding the mysteries of the brain. Each chapter offers science-backed ideas you can adapt to your own preferences and lifestyle. Part 2 ends with a brand-new and easy-to-follow twelve-week program to carry out the steps I suggest.

Part 3 takes a look at the challenges of diagnosing and treating brain diseases. What should you do if you notice the early signs? Are they symptoms of another health condition that mimics dementia? Why have our research and clinical trials failed so miserably in coming up with cures and drugs to treat neurodegenerative ailments? What treatments *are* available at all levels of severity? How can a spouse remain healthy while caring for a partner with dementia (caregivers have a much higher risk of developing the disease)? Dementia is a moving target; caring for someone with the disease can be one of the most challenging jobs ever undertaken. No one learns in formal schooling how to deal with a loved one whose brain is in irreversible decline. For some, the brain changes are slow and subtle, taking years or even more than a decade for symptoms to become pronounced; for others, it's sudden and rapid. Both circumstances can be difficult and unpredictable. In addition to covering evidence-based care that improves quality of life and makes the caregiving manageable, I'll also review highly treatable conditions caregivers should be on the lookout for that are often mistaken for Alzheimer's disease.

Finally, I'll look to the future, for this book ends on a high note. There is tremendous hope for the neurological conditions we still struggle with today (e.g., Alzheimer's, Parkinson's, depression, anxiety, panic disorders). I have no doubt that in the next ten to twenty years, we will be much further along when it comes to treating brain disorders. We may even have a successful therapy or preventive vaccine for Alzheimer's disease. Many of the advances may come from gene and stem cell therapy, along with deep brain stimulation, which is already being used for depression and obsessive compulsive disorder. We will also advance further technically, allowing a more minimally invasive approach to the brain. I'll explain what this all means for you and offer ideas to help prepare for this future. Many of the messages in this book are also directed at helping younger generations care for their own brain health since brain-related illnesses often start decades before symptoms show up. If I knew in my younger years what I know now, there are many things I would've done to take care of my own brain differently. You won't make the same mistakes I made.

I like an adage I once heard in Okinawa: "I want to live my life like an incandescent lightbulb. Burn brightly my entire life, and then one day suddenly go out." We want the same for our brains. We don't want the flickering of fluorescent lightbulbs that signal their impending demise. When we think of old age, we think of hospital beds and forgotten memories. Neither needs to happen, and your brain is the one organ that can get stronger as you age. There's nothing brainy about it—anyone can build a better brain at any age.

In a way, writing this book has been a selfish experience. I've had the privilege of going to specialists all over the world and getting their insights and action plans to keep my brain sharp and do all I can to prevent my brain from declining. Along the way, I've picked up on strategies to also be more productive, feel less overwhelmed, and generally navigate through life with ease and joy. I've been sharing this knowledge with everyone I hold near and dear to me. Now I want the same for you. Welcome to the Keep Sharp community.

Let's get started with a self-assessment.

Are You At Risk
for Brain Decline?

Over the past few years, I have spent a great deal of time distilling the best evidence-based brain research into guidance for you. It is based on formal and informal conversations with colleagues and other experts in the world of neuroscience and human performance. In order to make it most useful, I have created a list of questions that are highly relevant to your brain health and potential. No matter what in your life you are trying to improve, honest self-awareness is important, and answering these questions will help you do just that.

The list of twenty-four questions that follows will help you assess your risk factors for brain decline. These are mostly all modifiable risk factors, so don't panic if you answer yes to any of these questions. This is not meant to frighten you. (Remember: I don't believe that scare tactics work.) Some of these questions correlate with highly reversible symptoms of cognitive decline. Chronic sleep deprivation, for example, can lead to a staggering amount of memory loss that can even appear like the onset of dementia. Sleeping well is one of the easiest and most effective ways to improve all of your brain functions, as well as your ability to learn and remember new knowledge (it improves every system in the body). I underestimated the value of sleep for too long, taking great pride in my ability to function on a lack of it. Take it from me: That was a mistake. Luckily, this can be remedied with proper diagnosis and simply going to bed earlier and putting away your electronic devices and your to-do list. Some queries may seem unrelated, such as your level of education. For reasons I'll explain in this book, multiple studies now show that higher education might have protective effects in cognitive decline but not necessarily at slowing the decline once memory loss has started. In other words, people with more years of formal education (e.g., more college attendance and advanced degrees) or greater literacy have a lower risk for dementia than those with fewer years of formal education, but that doesn't matter as much if you start to develop dementia in the first place.

More than anything else, though, I want you to begin to see what kind of behavior plays a role in your brain health now and in the future. This is important. As a neurosurgeon, I know the satisfaction that comes with quick fixes, but you will see that some of these behavioral changes are not only effective but surgical in terms of rapid improvements. Knowing and understanding your daily habits will arm you with some personal data that can ultimately provide guidance for where you should be putting more effort—to rebuild and maintain a better brain. The questions are data driven insofar as they reflect scientific findings to date.

If you say *yes* to any of the questions below, that doesn't mean you'll receive a doomsday diagnosis now or in the future. Multiple factors, some of them not even included here because I wanted to keep this simple, are at play in the realm of cognition. Just as there are lifetime smokers who never get lung cancer, there will be people who live with many heightened risk factors for brain decline yet never experience it. Some of these risk factors are also debatable, and I will be transparent about that as well, alongside those debatable recommendations. Nonetheless, it's helpful to see all the potential risk factors for which there is good evidence and also the risk factors that researchers have been exploring and believe will be proven important in the future. I want to provide you with both the knowledge and the thinking that helped create that knowledge.

1. Do you suffer from any brain-related ailment now, or have you been diagnosed with mild cognitive impairment?

2. Do you avoid strenuous exercise?

3. Do you sit for most of the day?

4. Are you overweight or even obese?

5. Are you a woman?

6. Have you been diagnosed with cardiovascular disease?

7. Do you have any metabolic disorders such as high blood pressure, insulin resistance, diabetes, or high cholesterol?

8. Have you ever been diagnosed with an infection that can lead to chronic inflammation and can have neurological effects (e.g., Lyme disease, herpes, syphilis)?

9. Do you take certain medications with known possible brain effects, such as antidepressants, antianxiety drugs, blood pressure drugs, statins, proton pump inhibitors, or antihistamines?

10. Have you ever experienced a traumatic brain injury or suffered head trauma from an accident or playing an impact sport? Have you ever been diagnosed with a concussion?

11. Do you smoke or have a history of smoking?

12. Do you have a history of depression?

13. Do you lack social engagement with others?

14. Did your years of formal education end at high school or earlier?

15. Is your diet high in processed, sugary, fatty foods and low in whole grains, fish, nuts, olive oil, and fresh fruits and vegetables?

16. Do you live with chronic, unrelenting stress? (Everyone has stress. This is stress that seems to be constant or present more often than not and that you have trouble coping with.)

17. Do you have a history of alcohol abuse?

18. Do you suffer from a sleep disorder (e.g., insomnia, sleep apnea) or otherwise experience poor sleep on a regular basis?

19. Do you have hearing loss?

20. Does your day lack cognitive challenges in the form of learning something new or playing a game that requires a lot of thinking?

21. Does your job lack complex work with people in the form of persuasion, mentoring, instruction or supervision?

22. Are you over sixty-five years old?

23. Does Alzheimer's disease "run in your family," or have you been diagnosed with carrying the "Alzheimer's gene variant," APOE3 or APOE4, or both?

24. Do you care for someone who suffers from some form of dementia, Alzheimer's disease included?

If you answered yes to five or more questions, then your brain could be in decline or may be soon, and you can benefit tremendously from the information in this book. Even if you answered yes to only one or two questions, you can help optimize the health and performance of your brain for the better. Curious as to how these questions (and their answers) relate to your body's most mysterious organ? Read on to learn everything you want—and need—to know for a smarter, sharper, better-thinking you. Final reminder: This book isn't just about avoiding disease. It is about making your brain as sharp as it can be at any age.

Read on to learn everything you want—and need—to know for a smarter, sharper, better-thinking you. My hope is you can wind up like the couple who inspired me several years ago and who showed me what to aspire to when it comes to "old age." We all age and will one day live with an old brain, but that doesn't mean it has to lose its sharpness. Looks can be deceiving.

The husband was ninety-three years old and had been brought to the emergency room where I was on call. When my chief resident first told me about the patient, who was in serious neurological decline, his advanced age concerned me. I honestly thought he was too old to undergo an operation, should he need one. A little while later the CT scan showed a significant brain bleed that explained his symptoms.

I went to the family in the waiting room fully expecting them to tell me not to pursue an aggressive, risky operation. A spry woman who looked to be in her sixties was nervously pacing the room with several other family members sitting earnestly in chairs. I was shocked to learn she was his wife and they had just celebrated their seventieth wedding anniversary. "I am actually older than he is," she said. "I robbed the cradle." She was ninety-four years young in perfect health, took no medicines, and had driven her great-grandkids to school earlier that day. She shared that my patient was still an avid runner and worked part-time as an accountant. His sixty-three-year-old son said they kept him around because "he's such a whiz with numbers." His brain bleed occurred after he fell from his roof while he was blowing leaves up there. These nonagenarians were healthier than most of my patients, of any age.

Since I started medical school, there has always been a truism: We consider "physiological" age more than chronological age. At the family's request, I took the gentleman to the operating room for a craniotomy, which would fix the bleed. Before closing the dura, the outer layer of the brain, I took a few moments to closely inspect his brain, and what I saw surprised me. Given how active, cognitively intact, and sharp he was, I expected to see a large brain pulsating robustly and appearing healthy. But this looked like a ninety-three-year-old brain. It was more shriveled, sunken with deep wrinkles indicative of his age. Now, if this sounds disheartening to you, it should not. In fact, just the opposite.

Another truism in medicine is the following: Always treat the patient, not their test results. Yes, of course his brain had aged; he was ninety-three. But, the brain—perhaps more so than any other organ in the body—can reliably grow stronger throughout life and become more robust than in years past. I will never forget that experience. There seemed to be a total disconnect between the brain I was staring at and the man whose skull it inhabited.

He recovered quickly. When I visited him later on, recovering in the ICU, I asked him how the whole event affected him. He smiled and said, "The biggest lesson in all of this is no more trying to blow the leaves off the roof."

Final reminder: This book isn't just about avoiding disease. It is about making your brain as sharp as it can be at any age.

PART
1

THE BRAIN

MEET YOUR INNER BLACK BOX

In the seconds it takes for you to read this sentence, your brain will have fired off a miraculous number of electrical signals to keep you alive—breathing, moving, feeling, blinking, and thinking. Some of the information zipping through your billions of neurons are traveling faster than the speed of a race car. The human brain is a remarkable organ, an evolutionary marvel. It arguably houses more connections than there are stars in the known galaxy.[1] Scientists have said that the brain is the most complex thing we have ever discovered; one of the discoverers of DNA went so far as to call it "the last and grandest biological frontier. The brain," he said, "boggles the mind."[2]

Our brains sculpt who we are and the world we experience. It creates our everyday experiences, from those that bring us joy, wonder, and connection to fellow humans, to the complicated moments when we have to rely on our brains to make good decisions, plan, and prepare for the future. It even tells stories when we're sleeping in the form of dreams. And it knows how to adapt to environments, tell time, and form memories. It is quite likely the reservoir of our consciousness, though we are not entirely sure about that. (More on that later.) Neuroscientists have their work cut out for them because the brain continues to mystify as if it were a distant planet light-years away. It's arguably the most enigmatic 3.3 pounds of life. Researchers even found a new kind of neuron recently—the rosehip—and still don't know yet what it does. It seems to exist only in human brains but not in rodents, which may explain why so many mice brain studies never translate to humans. Our brains can be extraordinarily selfish and demanding as well. Of the total blood and oxygen that is produced in our bodies, the brain steals 20 percent of it, despite being only roughly 2.5 percent of your body weight. There can be no life without a brain.

Time to meet your inner black box.

CHAPTER 1

What Makes You *You*

Imagine the brain, that shiny mound of being, that mouse-gray
parliament of cells, that dream factory, that petit tyrant inside a
ball of bone, that huddle of neurons calling all the plays, that little
everywhere, that fickle pleasure dome, that wrinkled wardrobe of
selves stuffed into the skull like too many clothes into a gym bag.

DIANE ACKERMAN (FROM *AN ALCHEMY OF MIND*)

It was 1992 when I first saw a living human brain, a powerful and
life-changing experience for me. It was, and still is, hard for me to believe that so much of what we are, who we will become, and how we
interpret the world resides in that intricately woven bundle of tissue.
When I am describing a neurosurgery procedure, most people try to visualize what the human brain looks like, and they typically are a little off
base. For starters, it doesn't look like a dull and bland gray mass on the
outside, despite being referred to as gray matter. It is more pinkish with
whitish yellow patches and large blood vessels coursing on and through
it. It has deep crevasses, known as *sulci*, and mountainous peaks, known
as *gyri*. Deep fissures separate the brain into the various lobes in a surprisingly consistent way. During an operation, the brain pulsates gently
out of the skull's borders and looks very much alive. Consistency wise,
it is not so much rubbery as squishy, more like gelatin. It has always
amazed me how fragile the brain is despite its incredible function and

versatility. Once you see the brain, you very much want to protect it and take care of it.

To me, the brain has always been a bit mystical. Weighing in at a little over three pounds, it comprises all the circuitry we need to do just about everything. Think about that for a moment: It weighs less than most laptop computers, yet it can perform in a way that no computer can or will ever rival. In fact, the oft-cited metaphor of brains being like computers fails in oh-so-many ways. We may speak in terms of the brain's processing speed, its storage capacity, its circuitry, and its encodings and encryptions. But the brain doesn't have a fixed memory capacity that is waiting to be filled up, and it doesn't calculate in the manner a computer does. Even how we each see and perceive the world is an active interpretation and result of what we pay attention to and anticipate—not a passive receiving of inputs. It is true that our eyes see the world upside down. The brain then takes the input and converts it into a coherent image. In addition, the back of the eye, the retina, provides the brain with two-dimensional images from each eye, which the brain then converts into beautiful, textured three-dimensional images, providing depth perception. And we all have blind spots in our vision that our brain constantly fills in using constant data you probably didn't even realize you were collecting. No matter how sophisticated artificial intelligence becomes, there will always be some things the human brain can do that no computer can.

Compared to other mammals, our brain's size relative to the rest of our body is astonishingly large. Consider the brain of an elephant: It takes up 1/550 of the animal's total weight. Our brain, on the other hand, is about 1/40 of our body weight. But the feature that most sets us apart from all other species is our amazing ability to think in ways that reach far beyond mere survival. Fish, amphibians, reptiles, and birds, for instance, are assumed not to do much "thinking," at least in the way we conceive of it. But all animals concern themselves with the everyday business of eating, sleeping, reproducing, and surviving—automatic instinctual processes under the control of what's called the "reptilian brain." We have our own inner primitive reptilian brain that performs the same functions for us, and in fact it drives much of our behavior (perhaps more than we'd like to admit). It is the complexity and large size of our outer cerebral cortex that allows us to perform more sophisticated tasks than, say, cats and dogs. We can more successfully use language, acquire complex skills, create tools, and live in social groups thanks to that bark-like layer of our brain. *Cortex* means bark in Latin, and in this case, it is the outer layer of the brain, full of folds, ridges, and valleys. Because the brain folds back on itself over and over again, its surface area is far larger than you might guess—a little over two square feet on average, though exact calculations do vary (e.g., it would spread out over a page or two of a standard newspaper).[1] And probably somewhere deep in those crevasses is likely the seat of consciousness. Heady stuff!

The human brain contains an estimated (give or take) 100 billion brain cells, or neurons, and billions of nerve fibers (although nobody knows these numbers exactly for sure—because exact calculations are impossible as of yet).[2] These neurons are linked by trillions of connections called *synapses*. It is through these connections that we are able to think abstractly, feel angry or hungry, remember, rationalize, make decisions, be creative, form language, reminisce about the past, plan the future, hold moral convictions, communicate our intentions, contemplate complex stories, pass judgment, respond to nuanced social cues, coordinate dance moves, know which way is up or down, solve complex problems, tell a lie or a joke, walk on our tiptoes, notice a scent in the air, breathe, sense fear or danger, engage in passive-aggressive behavior, learn to build spaceships, sleep well at night and dream, express and experience deeply felt emotions such as love, analyze information and stimuli in an exceptionally sophisticated fashion, and so on. We can do many of these tasks at the same time, too. Perhaps you're reading this book, drinking a beverage, digesting your lunch, plotting when you'll get to your cluttered garage this year, thinking about your weekend plans ("in the back of your mind"), and breathing, among many other things.

Each part of the brain serves a special, defined purpose, and these parts link together to function in a coordinated manner. That last part is key to our new understanding of the brain. When I was in middle school, the brain was thought to be segmented by purpose—one area was for abstract thought, another for coloring within the lines, yet another for forming language. If you took high school biology, you may have heard the story of Phineas Gage, one of the most famous survivors of a serious brain injury. You may not know, however, just how much his unfortunate accident illuminated for scientists the inner workings of the brain at a time long before we had advanced techniques to measure, test, and examine brain functions. In 1848, the twenty-five-year-old Gage was working on the construction of a railroad in Cavendish, Vermont. One day, as he was packing explosive powder into a hole using a large iron rod measuring 43 inches long and 1¼ inches in diameter and weighing 13¼ pounds, the powder detonated. The rod shot upward into his face, penetrating Gage's left cheek. It traveled all the way through his head (and brain) and out the top. His left eye was blinded, but he didn't die and possibly didn't even lose consciousness or experience severe pain, telling the doctor who first attended to him, "Here is business enough for you." On the next page is a photo (known as a daguerreotype in early photographic technology) taken of Gage after recovering from the accident as he holds the offending tamping iron. This photo was only recently discovered and identified in 2009. To the right is a drawing made by Dr. John Harlow who treated him and recorded this sketch in his notes that became a publication of the Massachusetts Medical Society.[3]

Gage's personality, however, did not survive the blow intact. According to some accounts, he went from being a model gentleman to a mean, violent, unreliable person. The curious case of Phineas Gage was the first to demonstrate a link between trauma to certain regions of the brain and personality change. It had never been that clear-cut before. Keep in mind that in the 1800s, phrenologists still believed that measuring the size of bumps on a person's skull could be used to assess personality. Twelve years after the accident, Phineas Gage died at age thirty-six after experiencing a series of seizures. He has been written about in the medical literature ever since, becoming one of neuroscience's most famous patients. There was something else Phineas Gage taught us that is particularly important for this book. Some accounts of his life document a return of his more amiable nature closer to his death, which indicated the ability of the brain to heal and rehabilitate itself, even after significant trauma. This process of reestablishing networks and connections in areas of the brain damaged by the injury is what's called *neuroplasticity*, an important concept we'll be exploring. The brain is a lot less static than we thought in the past. It's alive, growing, learning, and changing—all throughout our lives. This dynamism offers hope for everyone looking to keep their mental faculties intact.

Although documentation on Gage's accident gave us a glimpse of the brain's complexity and the connection to behavior, it still took more than another century for us to understand that the brain's stunning power isn't simply due to its individual anatomical compartments. It's the circuitry

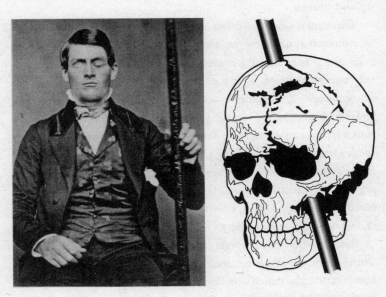

Photo of Phineas Gage and line drawing made by the town doctor, John M. Harlow, who helped treat him.

and communication *between those sections* that make up our complicated responses and behaviors. Many areas of the brain develop at different paces and in different stages of our life. For this reason, an adult solves problems differently and faster than a child does, an older person might struggle with motor skills such as walking and coordination in the dark, and the teenager might be a track-and-field star with perfect vision.

When most of us think about the brain, we probably think about the element of it that makes us, well, *us*. We ponder the mind—the part that includes our consciousness and is reflected by that quintessential inner voice or, as some would say, that monologic chatter we listen to all day long. It is your *you* that bosses you around all day, raises important as well as inane questions, beats you up emotionally on occasion, and makes life a series of decisions. I also have been mystified that every moment of jealousy, insecurity, and fear we have ever experienced lies within the caverns of the brain. And somehow the brain can take in data and create hope, joy, and pleasure.

The mind is what first drove me to study the brain. Oddly, however, we still don't really know precisely where consciousness resides in the brain or if it is even in the brain entirely. I find this to be a fundamentally important point. That state of being aware of oneself and one's surroundings—consciousness—on which everything else is predicated, remains elusive. Sure, I can tell you where in your brain rests the networks for processing sight, solving a math equation, knowing how to speak a language, walking, tying your shoelaces, and planning a vacation. But I cannot tell you exactly where your self-awareness comes from; it is probably the result of a confluence of factors throughout the brain—the outcome of metacognition, activities that involve multiple regions of the brain in their interconnectivity.

Getting to the brain is a highly orchestrated and meticulously planned journey. First, the skin is cut. Incidentally, it is the skin that contains pain fibers that must be dulled to perform brain surgery; the skull and the brain itself, that organ that innervates the entire body, has no sensory receptors of its own. It is why conducting brain surgery on an awake patient is an option (and probably why Phineas Gage felt little pain). The *dura mater* ("tough mother")—the layer covering the brain—has a few sensory fibers as well, but the brain itself does not. It is "so meta," as the kids say.

Once I've gotten inside someone's head (literally), I usually have a moment when I reflect on the fact that the brain can now be manipulated way too easily. After you have snuck into the castle (the skull), you have free rein. The brain floats in a bath of clear fluid and has no discernible smell. The brain offers nearly no resistance as you dissect, prod, probe, and cut. A patient could lose function of a limb if too much pressure is placed in one area or develop crippling dizziness from pressure in another section. A single snip could rob the patient's sense of smell, and a bigger snip could be blinding or worse. I've often wondered why the brain wouldn't put up more of a fight.

Knowing how vulnerable a brain can be on its exposure during surgery, I feel like a SWAT team member whenever I operate on one, or maybe I'm more like a highly trained thief. My goal is to get in, take what I need—say, a tumor, abscess, or aneurysm—and get the hell out without ever being detected. I want to disrupt the brain as little as possible.

Perhaps because it is encased in solid bone, the brain is often treated as a black box, viewed only in terms of its inputs and outputs without full knowledge of its internal workings. Impenetrable and indecipherable. And perhaps that is why the medical establishment simply resorted to the convenient adage that "what is good for the heart is also good for the brain." Truth is, though, that the saying became popular largely because both the heart and the brain have blood vessels. The brain, of course, is exponentially more intricate. What's more, the heart is a glorified pump, an engineering marvel for sure, but still a pump that can now be replicated in an engineer's laboratory. There is no true metaphor for the brain. If you become brain dead due to some horrible head injury, there is no replacement. It is the command central for not just our body but for our existence. Despite how much we have mapped it, probed it, infused it with chemicals, we are still not exactly sure what makes it tick or slows its tick. This has no doubt played into our frustrations in understanding and treating neurodegenerative decline and complex disease processes and disorders of the brain, from autism to Alzheimer's.

Now here's the silver lining: We may never know all the mysterious perplexities of the human brain and come to control it like my parents can an automobile, and that is okay. Maybe we are not supposed to know where consciousness resides or how our personal perceptions and perspectives are neuronally born. No, we can't touch our brains the way we can our skin or nose, but we know it's there, just like the air we breathe and the wind we feel on our face. We also know it's home to another bewildering marvel we cannot see, touch, or feel but immediately associate with the brain: memory—the process of remembering—but it's much more than that, as you are about to learn. It is what makes us uniquely human, and it's the first pillar of having a sharp, fast-thinking, resilient brain.

Brainy Facts

- The typical human brain comprises about 2 to 2.5 percent of the body's total weight but uses 20 percent of its total energy and oxygen intake.
- Your brain is roughly 73 percent water (same for your heart), and that is why it takes only 2 percent dehydration to affect your attention, memory and other cognitive skills, so drinking just a few ounces of water can reverse that.
- Your brain weighs a little over three pounds. Sixty percent of the dry weight is fat, making the brain the fattiest organ in the body.
- All brain cells are not alike. There are many different types of neurons in the brain, each serving an important function.
- The brain is the last organ to mature. As any parent can attest, children's and teenagers' brains are not fully formed, which is why they take to risky behaviors and can have a harder

time regulating their emotions. It isn't until about the age of twenty-five that the human brain reaches full maturity.

- Brain information can travel faster than some race cars, up to more than 250 miles per hour.
- Your brain generates enough electricity to power a low-wattage LED light.
- The average brain is believed to generate tens of thousands of thoughts per day, give or take.
- Every minute, 750 to 1,000 milliliters of blood flow through the brain. This is enough to fill a wine bottle and then some. Every minute!
- Your brain can process a visual image in less time than it takes for you to blink.
- The hippocampus, the part of the brain considered the memory center, has been documented to be significantly larger in people whose jobs have high cognitive demands, compared to the average person. London cab drivers, for instance, get a mental workout while navigating London's 25,000 streets. However, those memory centers may be getting smaller because of GPS.
- Your brain starts slowing down by the surprisingly young age of twenty-four, right before maximum maturity, but it peaks for different cognitive skills at different ages. No matter how old you are, you're likely still getting better at some things. An extreme case is vocabulary skills, which may peak as late as the early seventies![4]

THE ESSENCE OF MEMORY, THINKING, AND HIGH MENTAL FUNCTIONING

Memory, as the ancient Greek dramatist Aeschylus said, is the mother of all wisdom. But it's also the mother of everything about us. The smell of your grandma's cooking, the sound of your child's voice, the image of your late father's face, the thrills of a vacation you took twenty years ago: These are the memories that form our ongoing experiences of life and give us a sense of self and identity. Memories are what make us feel alive, capable, and valuable. They also help us feel comfortable with certain people and surroundings, connect the past with the present, and yield a framework for the future. Even bad memories can be useful, as they help us to avoid certain situations and inform better decision making.

Memory is the most commonly recognized cognitive function, a higher brain function. In addition to memory, cognition includes attention, writing, reading, abstract thinking, decision making, problem solving, and performing everyday tasks like navigating your way while driving, figuring out the tip at a restaurant, appreciating the benefits or harmful effects of the food you eat, or admiring the works of different artists. Memory is the cornerstone of all learning, for it is where we store knowledge and process that knowledge. Our memory must decide what information is worth keeping and where it fits in relation to previous knowledge that we have already stored. What we store in our memories helps us process new situations.

Many of us, however, mistake memory for "memorizing." We view memory as a warehouse where we keep our knowledge when we are not using it, but that metaphor is not correct because memory is not static like a physical building. Our memories are constantly changing as we take in fresh information and interpret it. From your brain's perspective, new information and experiences in the future can change the memories of our past. Consider this in evolutionary terms: Being able to recall all the details of a particular event is not necessarily a survival advantage. The function of our memory is more about helping to build and maintain a cohesive life narrative that fits with who we are while also constantly changing with new experiences. This dynamism is partly why it's also true that our memories are not an accurate, objective record of the past. They can be contaminated or changed fairly easily, even in people who have no problems with their memory. Years ago, I did a story about Bugs

Bunny and Disney World. It was based on research by psychology professor Elizabeth Loftus, in which she presented advertisements featuring the characters to visitors at a Disney theme park. Some of the ads showed Bugs Bunny, and the people who saw those ads were often convinced they had in fact met Bugs Bunny in the Disney park and even shook hands with him. They would sometimes describe a carrot in his mouth, his floppy ears, and things he may have said, like, "What's up, Doc?" The problem is that Bugs Bunny is a Warner Bros. character and would never be seen in a Disney park. Loftus demonstrated just how easily memories can be implanted or manipulated.

Now consider what happens as you read an article in a magazine, in a newspaper, or online. As you digest the new information, you're using information you've already got tucked in your memory. The new information also evokes certain ingrained beliefs, values, and ideas that are unique to you and help interpret the information, make sense of it, fit it into your worldview, and then decide whether you will retain it (while altering previously stored information) or let it be forgotten. Thus, as you read the article, your memory actually changes by both adding new information and finding a new place to put that information. At the same time, you're giving yourself a different way to link the new information with older, now slightly modified information. It's complicated, and probably not at all how you have previously thought about your memory. But it is important to know that memory is fundamentally a learning process—the result of constantly interpreting and analyzing incoming information. And every time you use your memory, you change it. This is important. When we talk about improving or preserving memory, we need to first understand what it is, and what it represents to any given person.

We tend to worry about our ability to remember names or where we put our keys, but we should also worry about the memory we need to be a great performer in whatever role we fill as a professional, parent, sibling, friend, innovator, mentor, and so on. Whether we're talking about the kind of memory we need to remain cognitively intact for the duration of our life and avoid dementia, or whether we're referring to the memory we need to achieve peak performance in our everyday goals and responsibilities, we're talking about the same thing—the same memory. The reason I am describing this in such detail is that the more you understand your memory, the more inspired you'll be to improve it.

It wasn't that long ago that neuroscientists described memory using metaphors like a filing cabinet that stores individual memory files. But today we know that memory cannot be described in such concrete terms. It is far more complex and dynamic. We also now know that memory is not really confined or generated in one particular location of the brain. It is a brain-wide active collaboration that involves virtually every part of it when running full throttle. That is why it makes sense that new research is showing promise for the ability to tune our memories. Because memory calls on an expansively distributed network and coordinates those interactions through slow-frequency, thrumming rhythms called theta waves, neuroscientists are finding ways to stimulate key regions in the brain with noninvasive electric currents to physically synchronize neural circuits, akin to an orchestra conductor tuning the strings section to the wind. This kind of research and resulting potential therapy is in its infancy, but the belief is that one day we may be able to tune a seventy-year-old's memory into that of someone decades younger.

If I asked you to recall what you had for dinner last night, an image might come to mind. Perhaps it was a plate of chicken marsala or a bowl of chili. That memory was not sitting in some neural alleyway waiting to be retrieved. The mental picture of your dinner was the outcome of an astoundingly intricate choreography of processes scattered throughout the brain that involved multiple neural networks. Construction of a memory is about reassembling different memory "snapshots" or impressions from a lattice-like pattern of cells found throughout the brain. In other words, your memory is not a single system—it's comprised of a network of systems, each playing a unique role in creating, storing, and recalling. When your brain processes information normally, all of these systems work together in synchronicity to provide cohesive thought. Single memories therefore are the result of a complex construction. Think of your favorite pet. Let's say it's a dog named Bosco. When your brain pictures the dog, it's not just grabbing a memory of what Bosco looks like from one place. It retrieves the dog's name, appearance, behavior, and sound of his bark. Your feelings toward him also participate. Each part of the memory of what Bosco encompasses comes from a different region of the brain, so your comprehensive image of Bosco is actively reconstructed from many areas. Scientists who study the brain are only beginning to understand how the parts are marshalled into a coherent whole. You can think of it like this: When you recall a memory, it is like assembling a giant jigsaw puzzle from a few small pieces to get it started. As the pieces come together, link, and define an image, they begin to tell a story, convey a picture, or share knowledge. The puzzle becomes grander and larger, showing more and more of its meaning. By the time you place the last piece, you have gathered the information to complete a full "memory." Given this analogy, you can see that in order for memory to work properly, the right pieces of the puzzle have to be present and attached together appropriately, which is akin to integrating information from all these different parts of the brain into something that makes sense. If pieces are missing or are not joined together as designed, the memory won't come together perfectly. There will be gaps, holes, and an undefined result.

Music is an illustrative example. If you want to sing a song, you must first retrieve the words and be able to say them. That typically involves the left side of the brain, specifically the temporal lobe. Singing those words requires more than just saying them: You have to engage the right parietal and temporal lobes, which handle nonverbal memory such as pitch and tone. All of this information must move to and from the right and left sides of the brain to sync up and integrate data. If you want to add a rhythm or beat to the music, that usually comes from the back of the brain, known as the *cerebellum*. You get the idea. Watching the brain of someone who is singing a song in a functional magnetic resonance imaging (fMRI) scanner is like watching a light show on a clear night sky. And yet we know people with even advanced dementia who can still sing songs from their childhood without a problem. Collectively, disparate places in their brains are still able to coordinate and work together, even if discrete parts of the memory system are starting to fail.

The same elaborate process takes place when you perform what would otherwise seem to be a single action, such as driving a car. Your memory of how to operate the vehicle comes from one set of brain cells; the memory of how to navigate the streets to get to your destination springs from another set of neurons; the memory of driving rules and following street signs originates from another family of brain cells; and the thoughts and feelings you have about the driving experience itself, including any close calls with other cars, come from yet another group of cells. You do not have conscious awareness of all these separate mental plays and cognitive neural firings, yet they somehow work together in beautiful harmony to synthesize your overall experience. In fact, we don't even know the real difference between how we remember and how we think. But, we do know they are strongly intertwined. That is why truly improving memory can never simply be about using memory hacks or tricks, although they can be helpful in strengthening certain components of memory. Here's the bottom line: To improve and preserve memory at the cognitive level, you have to work on all functions of your brain.

Scientists haven't sorted out the precise physiology behind how the brain thinks, organizes memories, and recalls information, but they have offered enough of a working knowledge to state a few reliable facts about this amazing feat.

It helps to consider memory building in three phases: encoding, storage, and retrieval.

Building a Memory (Encoding)

Creating a memory starts with encoding, which begins with your perception of an experience using your senses. Think about your memory of meeting someone you fell in love with, perhaps even married. In that first meeting, your eyes, ears, and nose took note of that person's physical features, voice sounds, and personal scent. Maybe you also touched the person. All of these separate sensations traveled to the hippocampus, the area of the brain that integrates these perceptions or impressions as they were happening into one single experience—in this case, the experience of the individual.

While the function of memory is facilitated in areas throughout the brain, the hippocampus is your brain's memory center. (Studies show that as your hippocampus shrinks, so does your memory; studies also show that a higher waist-to-hip ratio—ahem, carrying extra weight—equates with a smaller hippocampus. More on this later.) With the help of the brain's frontal cortex, your hippocampus takes the helm to analyze these various sensory inputs and evaluate whether they are worth remembering. Now, it's important to understand how memory and learning are occurring at a biochemical level, which will help you grasp why the strategies I suggest will work for you. All of the analyzing and filtering of your perceptions occur using the brain's language of electricity and chemical messengers. As you already know, nerve cells connect with other cells at an endpoint called the synapse. Here, electrical pulses carrying messages jump across super small spaces or "gaps" between cells, which triggers the release of chemical messengers aptly named neurotransmitters. Examples of common neurotransmitters are dopamine, norepinephrine, and epinephrine. When they move across these gaps between cells, they attach themselves to neighboring cells. A typical brain has trillions of synapses. The segments of the brain cells that receive these electric impulses are called *dendrites*, which literally means "treelike" because they are short branched extensions of a nerve cell that reach out to nearby brain cells.

The attachments between brain cells are incredibly dynamic in nature. In other words, they are not fixed like a line of cable. They change and grow (or shrink) continually. Working together in a network, brain cells organize themselves into specialized groups to serve in different kinds of information processing. When one brain cell sends signals to another, the synapse between the two strengthens. The more often a particular signal is sent between them, the stronger the connection grows.

That is why "practice makes perfect." Every time you experience something new, your brain slightly rewires to accommodate that new experience. Novel experiences and learning cause new dendrites to form, whereas repeated behavior and learning cause existing dendrites to become more entrenched. Both are important, of course. The creation of new dendrites, even weak ones, is called *plasticity*. It is this plasticity that can help your brain rewire itself if it is ever damaged. It is also the core ingredient for resilience, vital for building a better brain (see chapter 3). So, as you navigate the world and learn new things, changes happen at the synapses and dendrites—more connections are generated, while some may be weakened. The brain perpetually organizes and reorganizes itself in response to your experiences, your education, the challenges you face, and the memories you make.

These neural changes are reinforced with use. As you learn new information and practice new skills, the brain builds intricate circuits of knowledge and memory (hence the saying "what wires together fires together"). If you play Beethoven's Moonlight Sonata on the piano over and over, for example, the repeated firing of certain brain cells in a certain order makes it easier to replicate this firing later. The result is that you get more adept at playing the piece effortlessly. You can play it without even thinking about it note for note, measure by measure. Practice it repeatedly and long enough, and you will eventually be able to play it flawlessly "from memory." If you stop practicing for several weeks, though, and then try to play the piece again, you may not be able to play it as perfectly as before. Your brain has already begun to "forget" what you once knew so well. The dendrites that were so well defined start to wither away a bit fairly quickly. Luckily, it is not difficult to read the notes even years later and build up those neural connections once again.

There is a caveat, however, to all this memory making. You have to pay attention to properly encode a memory. Do you need to read that again? Put simply, you must have an awareness of what you're experiencing. Because you cannot pay attention to everything you encounter, a lot of potential stimuli is automatically filtered out. In reality, only select stimuli reach your conscious awareness. If your brain remembered every single thing it noticed, its memory system would be overwhelmed to the point you'd find basic functioning difficult. What scientists are not sure about is whether stimuli are screened out after the brain processes its significance or during the sensory input stage. How you pay attention to incoming data, however, may be the most important factor in how much of that information you remember.

I should point out that forgetting does have significant value. As I mentioned, if you remembered everything that comes into your brain, your brain would not work properly and your ability to creatively think and imagine would be diminished. Everyday life would be difficult; sure, you'd be able to recall long lists and cite elegiac love poems, but you'd struggle to grasp abstract concepts and even to recognize faces. There's a group of neurons that are charged with helping the brain to forget, and that are most active at night during sleep when the brain is reorganizing itself and preparing for the next day of incoming information. Scientists discovered these "forgetting" neurons in 2019, which helps us further understand the importance of sleep—and the merits of forgetting. It's a beautiful paradox: In order to remember, we have to forget to some degree.

Short- versus Long-Term Memory (Storage)

It's common knowledge that our memories work on two different levels: short-term memory and long-term memory. But even before an experience can become part of our short-term memory, which includes what you focus on in the moment—what holds your attention—there is a sensory stage that lasts a fraction of a second. During this initial stage, your perception of an experience gets logged in your brain as you register incoming information—what you see, feel, and hear. Sensory memory allows that perception to remain after the stimulation is over, though only momentarily. Then the sensation moves into short-term memory.

Most of us can hold only about seven items of information in short-term memory at any given time, such as a list of seven grocery items or a seven-digit phone number. You may be able to increase this capacity a little through various memory tricks or strategies. For example, a ten-digit number such as 6224751288 may be too long to remember all at once. But broken up into orderly blocks, as in a hyphenated telephone number, 622-475-1288 you'll more easily store that in your short-term memory and be able to recall it (your Social Security number is hyphenated and therefore easier to remember). Repeating the number to yourself also helps pack the information into short-term memory. To learn information so that you can retain and recall it, you must transfer it from short-term to long-term memory. Short-term memory is closely linked to the function of your hippocampus, while long-term memory is closely linked to functions of the outer layer of the brain, your cortex (see image below).

Long-term memory includes all the information that you really know and can recall. In many ways, it becomes part of you. It's how you remember events from last week, last year, or your childhood. Once informa-

Short-Term Memory: Hippocampus
Long-Term Memory: Cortex

Cortex · Hippocampus

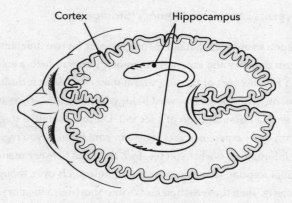

Short- vs. Long-Term Memory Areas of Your Brain

tion becomes part of your long-term memory, you have access to it for a long time. Unlike sensory and short-term memory, which are limited and decay rapidly, long-term memory allows us to recall functions to store unlimited amounts of information indefinitely. Certain things can interrupt the process of moving a memory from short term to long term, however. Alcohol, for example, puts a glitch in the process. For someone who is intoxicated, the encoding into long-term memory often does not occur very well, or at all. And that is why, days later, someone may have a hard time remembering something that was so vivid earlier when the memory was still in short-term storage. In these cases, they can't retrieve the memory from the long-term bucket because it was never there in the first place. Sleep deprivation can also disrupt the movement of memories from short to long term. During sleep, your body consolidates and transfers your short-term memories to long-term memories—the kind you'll have for much of the rest of your life.

Retrieval

Of course, none of this works without retrieval. When you call up a memory, you first fetch the information at an unconscious level and willfully drop it into your conscious mind. Most people see themselves as having either a "good" or a "bad" memory, but the truth is we each are fairly adept at remembering some types of things and not so great at remembering others. If you struggle with remembering, say, people's names and you're not suffering from a physical disease or dementia, it's usually not the failing of your entire memory system. It could be a lack of attention at the time you were being introduced and first heard the person's name. It could also be an inefficient retrieval system. In those cases, people often feel like the name is "on the tip of their tongue." Sometimes that can easily be rectified by sharpening your memory skills for that particular weakness, encoding or retrieval. Many memory champions started off believing they had poor memory until they spent time practicing techniques focusing on a very specific component of memory.

For some people, however, memory problems do tend to increase as they get older. Memory speed and accuracy naturally begin to slip in our twenties, especially our working memory that holds information temporarily in the mind so we can get through the day and make good decisions. But as I reiterate throughout this book, memory problems are not inevitable with age. There are things we can do to maintain, enhance, and sharpen our abilities to remember, retain, and retrieve that information as long as we live. Now, let's turn to some of the terminology you'll need going forward. How is cognitive decline defined? What is considered normal and abnormal? Is it reversible?

CHAPTER 2

Cognitive Decline—Redefined

You'd better start mixing toothpaste with your
shampoo. You're getting a cavity in your brain.

ARCHIE BUNKER (*ALL IN THE FAMILY*, 1971)

When my friend Sarah described to me her mother's experience with cognitive decline over several decades that accelerated after she retired at the age of sixty-two, I thought of my grandfather. I was immediately reminded of how agonizing it can be to watch someone wither away mentally—and emotionally. For many people, the downward slope is slow and steady, like a long-drawn-out illness, while for others it's fierce and fast, like a traumatic accident.

The first questions that often enter a family member's mind when a loved one seems to be struggling with cognition are the following: *When did it start? What caused it? What can I do to help?* That's what Sarah asked herself when she noticed something was not right with her mother's brain. The best Sarah could describe her mother's short-term memory was that it seemed to have a "serious malfunction." It is interesting that we detail most medical maladies with words like *painful, blockage, tumor,* or *swelling,* but with dementia, we often resort to mechanical explanations, as Sarah had. One of the first signs with Sarah's mother was that she often misspoke her grandson's name, calling him Conner instead of Colin. Over time, she stopped socializing and engaging in normal everyday activities like cooking, tidying up the house, and tending to her general hygiene. Although she had a history of mild depression, her anxiety and moodiness reached all-time highs, and increasingly she had no filter— she'd make hurtful, rude, inappropriate comments, and sometimes use abrupt foul language. After retiring, she became mostly housebound by choice, steadily distancing herself from friends. She was more frequently choosing to sit in front of the television rather than read books, go for long walks, or retreat to the beach as she had enjoyed throughout her life. Sarah's father, who still worked full-time, had to take on all the house-work and the bills. As I shared Sarah and her mom's story with the ex-perts I interviewed, they all said it was a familiar pattern of symptoms. The progression is often very similar, starting with little slips and transi-tioning into more and more withdrawal.

When Sarah's mother began to get lost while driving or routinely leave the car in parking lots because she couldn't find it after shopping (or thought she had walked), the keys were taken away. Her mood also changed. Her mother was always a bit depressive, which made Sarah wonder how much a lifelong bout with untreated depression contrib-uted to her mother's mental demise. Or was it the daily chardonnay habit that did it? The lack of sufficient exercise? The nutritional deficiencies from an eating disorder that began in her youth and never really went away—even with treatment? How much did not engaging in social activ-ities, hobbies, and challenging work contribute to the acceleration of the disease? These are the questions millions of families ask themselves, and there are often few satisfactory answers.

Sarah's story brings light to the fact that we often don't and can't know what triggers cognitive decline in the first place or what propels it over time. Multiple forces are likely in play, as there is no single culprit. Many theories exist, but we have no definitive answer yet. It is becoming abun-dantly clear, though, that the decline starts years, if not decades, before any symptoms emerge. This is a crucial concept: A thirty-year-old can be on the road to Alzheimer's disease but not know it. People often don't think or worry about dementia until after they turn fifty, which is why it's

so important that younger generations heed the message and start thinking of habits that can help prevent decline.

While we have made so much progress in medicine, more than a century after the first description of Alzheimer's disease by a German psychiatrist and neuropathologist whose surname is forever attached to the
condition, researchers still cannot identify the precise cause or causes. It
is a reminder that we humans are exceedingly complex organisms. It also
means that what causes grave cognitive decline in person A will not cause
it in person B, or C, or D, and so on. Sarah's mother and my grandfather
both received the diagnosis of Alzheimer's disease, but probably for very
different reasons. It's like cancer: What causes breast or colon cancer in
one individual will not always be the same for anyone else. There are
myriad pathways to any particular type of cancer, and the same is true
with dementia. Despite that, as we take a deeper look at the data, we realize there are still some excellent insights and strategies to reduce our risk
for dementia.

To better understand these strategies, it is worth once again examining current theories on what is happening in the brain of someone with
Alzheimer's. As many of you have probably read, the amyloid hypothesis has led the charge in recent decades. Amyloid, or more precisely
beta-amyloid, are plaques of sticky protein that accumulate in the brain
and destroy those essential synapses that allow brain cells to communicate. The problem is that treatments based on this hypothesis, including medicines to extinguish those plaques, have largely failed in clinical
trials. When Merck ended its study on a once-promising Alzheimer's
disease drug in 2017, Mayo Clinic neurologist Dr. David Knopman told
Bloomberg that "[r]emoving amyloid once people have established dementia is closing the barn door after the cows have left."[1]

The progression of the disease, it turns out, is a lot more complex than
any single culprit. Researchers have also investigated whether cognitive
decline is simply an acceleration of normal aging or a degenerative disease of specific brain pathways. To that end, recent research has focused
on possible triggers: infections, injury, nutrient deficiency, prolonged

metabolic dysfunction, exposure to harmful chemicals—all of which can stimulate an immune response and inflammatory reaction that damages the brain. That brings us to inflammation, a keyword you're going to read about over and over again. As you'll learn shortly, inflammation is a common thread in all of the theories about brain decline, not to mention most other types of illness. Once you understand this concept, several of the strategies to lower your risk will make more sense.

I'll slow this down for a moment by taking a quick tour of the most common and possible causes of cognitive decline beyond normal or even accelerated aging. As you read through this list, you will see just how much genetics, lifestyle, and environmental factors contribute to the problem.

EIGHT (POTENTIAL) WAYS THE BRAIN BEGINS TO BREAK

Many of the factors outlined in this section can be part of the problem, some more influential than others depending on individual risk factors.

The Amyloid Cascade Hypothesis (ACH)

When Dr. Aloysius Alzheimer first described "a peculiar disease" in a fifty-one-year-old woman who had profound memory loss, bizarre behavior, and unexplained psychological changes, he would go down in history as the original documenter of the haunting disease that now bears his name. During the autopsy of her brain, he saw dramatic shrinkage and abnormal deposits in and around nerve cells that he called "senile plaques" in his 1907 report, which would later be recognized as containing beta-amyloid. Today, more than one hundred years later, these amyloid plaques (plaques), along with neurofibrillary tangles (tangles), remain the hallmarks of Alzheimer's disease. Picture it like this: in Alzheimer's disease, amyloid plaques that accumulate between nerve cells and tangles, primarily consisting of tau protein, are twisted insoluble fibers found *inside* the brain's cells. (Beta-amyloid was discovered in 1984, with tau identified two years later. Tau protein is a microscopic component of brain cells that is essential to their stability and survival; more on tau shortly.)

Here's the tricky part: We need beta-amyloid and tau in our brains. Healthy versions of these proteins are part of healthy brain biology: They help supply food to brain cells and make sure important chemicals move freely between those cells. It's when beta-amyloid and tau become damaged, misfolding into sticky clumps, that problems arise. The amyloid fibrils turn rogue when they morph into watertight rope-like structures containing proteins that interlock like the teeth of a zipper. These tight molecular zippers become sealed and difficult to pry apart, gumming up to form dangerous plaques. According to the amyloid cascade hypothesis, it is the accumulation of plaques around brain cells that causes Alzheimer's disease, even though scientists aren't sure how or why that happens. Drugs to reduce beta-amyloid in the human brain have not succeeded as expected. A string of clinical failures based on this hypothesis has punched lots of holes into the idea that beta-amyloid tells the whole story. Some people whose brain is riddled with plaques show no signs of cognitive decline. These brains are often found to be filled with plaques on autopsy, and yet the patients died cognitively intact. While this can be due to what's called cognitive reserve, a topic I'll cover at length, in truth we don't really know whether the plaques are an effect rather than a cause of Alzheimer's.

In the Alzheimer's world, a "unicorn" is when the brain autopsy of someone with dementia only shows damage from plaques and tangles. The point is that rarely does a diseased brain show only one form of damage: lots of changes in an aging brain can result in a diagnosis of Alzheimer's disease. The complexity of this disease has forced scientists to rethink their entire approach and search for a cure. There probably won't be a universal solution. People likely have a mix of different dementias for which they need a mix of different treatments.

Genetics can also be a factor. Certain gene abnormalities, such as mutations in the genes that code for amyloid protein—namely, the amyloid precursor protein (APP) gene and presenilin 1 and presenilin 2 genes—can increase beta-amyloid production and account for the early-onset

Alzheimer's that affects many members of the families that carry these mutations. In one particular cluster of cases in a South American family, for example, many family members showed cognitive impairment around age forty-seven, progressing to dementia by about age fifty-one, and they died around age sixty. Scientists have been studying mutation clans throughout the world where the disease runs heavily in families; and sometimes within these clans are individuals who have the genetic profile to develop early-onset Alzheimer's and they are somehow protected from that fate thanks to other rare mutations. These lucky people's brains show neurological features of the disease but there's no outward signs of cognitive decline.

The hope is that by understanding the natural history of the disease with strong genetic roots, scientists can develop new drug or gene therapies, including for those who do not carry Alzheimer's-causing mutations yet still develop dementia. These amyloid-related genes and their products are highly complicated with many functions beyond just the brain's neurons; they can be difficult to study, but the more we learn about how they operate and result in disease (or not), the faster we can arrive at solutions. You have probably heard about the APOE genes linked to Alzheimer's; these are one of many sets of genes that could be linked to an increased (or decreased) risk of late-onset Alzheimer's (after age sixty-five). I'll go into detail about these genes in a later chapter.

While early-onset Alzheimer's is more likely to be influenced by genetics, genes may play some role later in life as well. What makes the body especially vulnerable as we get older is that the repair system for fixing DNA mutations becomes less efficient. For example, the dry molecular amyloid "zipper" I described may start with a single kink in the chain of amino acids. As we get older, those kinks build up because repair enzymes can no longer keep up. This is similar to what happens in cancer: DNA repair weakens as we age, making us more susceptible to cancer when genetic mutations build up and trigger cancerous growths. Scientists are trying to understand these zippers to unlock the cascade that results in Alzheimer's disease. One international team, led by UCLA's Professor David Eisenberg, is hoping these sorts of insights will eventually lead to new therapies.

Normal brain cell with healthy tau inside and beta-amyloid outside.

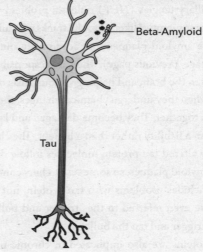

Beta-Amyloid

Tau

Diseased brain cell with tangled tau inside and amyloid plaques outside.

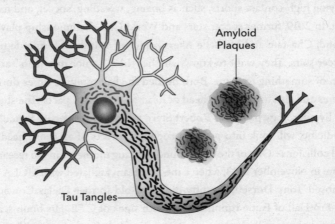

Amyloid Plaques

Tau Tangles

Tau and Tangles

Neurofibrillary tangles (NFT) reflect a problem with tau protein. Tau proteins are sometimes likened to the tracks of a railroad inside brain cells (versus the amyloid plaques that accumulate outside brain cells—see bottom image, previous page). They are responsible for both stabilizing nerve cells in the brain and helping various areas of the brain communicate. But when they undergo chemical changes, they no longer help hold nerve cells together. They become damaged and knotted up, which then makes them a liability rather than a benefit. The clumping and spread of chemically altered tau protein molecules follow different patterns from those of amyloid plaques, so some researchers continue to look for a theory that includes problems with tau protein, not just amyloid. Recent papers have even referred to the "trigger and bullet" theory—amyloid being the trigger and tau the bullet.[2]

Tau proteins are also implicated in chronic traumatic encephalopathy (CTE), a degenerative brain disease linked to repeated blows to the head and associated with behavior problems, depression, memory loss, and dementia. CTE is especially prevalent in professional athletes playing high-contact sports, such as boxing, wrestling, soccer, and football. In 2019, former soccer stars and World Cup championship players Brandi Chastain and Michelle Akers launched a study of former female soccer stars. They want to know whether their "senior moments" are a sign of something to come. Both headed the ball multiple times during games and also had head-to-head or head-to-ground impacts. The study, led by neurology professor Robert Stern at Boston University School of Medicine, will look into possible cognitive effects of all those headers and collisions.[3] One of the first groundbreaking moments in tau research came in November 2013, when a medical team affiliated with UCLA diagnosed Tony Dorsett, the fifty-nine-year-old former Dallas Cowboys All-Pro Hall of Fame running back, with signs of CTE. His brain scans showed abnormally high concentrations of tau. It was among the first cases that a living person was diagnosed with degenerative CTE.

Prions are increasingly becoming part of the narrative around plaques and tangles. Prions are another type of protein found in the brain that can trigger other proteins (like beta-amyloid and tau) to fold abnormally. A few diseases are attributed to prions, which are associated with infections and are universally fatal. The most common form of prion disease in humans is Creutzfeldt-Jakob disease (a.k.a. "mad cow disease") from infected meat products. Some researchers are looking into whether prion-like forms of both amyloid and tau spread through the brain, forcing normal proteins to become misfolded and knotted up, setting the stage for Alzheimer's disease.

Blood Flow

Plaques, and sometimes tangles, are known to occur more frequently and severely in people with advanced vascular disease, which is a class of diseases that affect the blood vessels (arteries and veins). This suggests that blood flow abnormalities in the brain may be important in the development of Alzheimer's disease. Decreased blood flow to the brain, a condition called hypoperfusion, has long been associated as a precursor to the buildup of plaques and tangles. It is likely that changes in blood flow to the brain create a crisis among neurons and their support staff cells called glia, leading to degeneration of these cells and subsequent cognitive impairment. Remember that the brain is a highly vascular organ; it demands a lot from the circulatory system to continually deliver nutrients and oxygen. Any factor—from smoking to high cholesterol levels—that affects the blood flow system in the brain has a significant impact on its function and risk for decline.

In addition, the vascular hypothesis of Alzheimer's disease may explain why people who have a history of high blood pressure or have had a stroke are more vulnerable to developing the disease. High blood pressure can cause microscopic damage in the arteries leading to the brain, which can further reduce blood flow and oxygenation. Brain cells need energy in the form of glucose and oxygen. When that energy to the active brain is compromised by lack of adequate blood flow, trouble looms. Recent research has also shown that blood flow to the brain is decreased when the blood-brain barrier, a semipermeable barrier in the brain's capillaries, breaks down.[4] Because the brain is so precious, it is protected not only by the skull and a bath of cerebrospinal fluid; the blood-brain barrier also effectively walls off the brain from the body's blood supply. When working properly, this wall lets oxygen, glucose, and other necessary substances across the barrier but stops larger, sometimes toxic, molecules from getting into the brain. Gaps can form in this barrier, however, thereby letting harmful molecules enter the brain and accumulate. The result is a gradual swelling of the brain, which increases the pressure inside the cranium and inhibits blood flow to the brain. And once again, with less oxygenated blood reaching the brain, the crisis in neurons and glia is ignited. In turn, this causes more brain swelling, lesions, and the formation of beta-amyloid plaques and tau tangles. Recent studies have shown the hippocampus is especially vulnerable to this "leaky blood-brain barrier" condition, and as it loses it protective barrier, toxic substances from the blood vessels can penetrate the neurons and worsen an individual's memory loss and cognitive impairment.[5]

Metabolic Disorders

Another big risk factor for dementia is the broad category of metabolic disorders. Nearly 35 percent of all U.S. adults and 50 percent of those sixty years of age or older are estimated to have what's called metabolic syndrome, a combination of health conditions you don't want to have, such as obesity, high blood pressure, insulin resistance, type 2 diabetes, or a poor lipid profile (too much bad cholesterol, not enough good cholesterol).[6] Since 2005, researchers have been finding correlations between diabetes and risk for Alzheimer's disease, especially when the diabetes is not controlled and a person suffers from chronic high blood sugar.[7] Some have gone so far as to refer to Alzheimer's disease as "type 3 diabetes," because the disease often involves a disrupted relationship with insulin, the metabolic hormone involved in both types 1 and 2 diabetes. Insulin is the hormone needed to deliver sugar (glucose) into cells for use. Without insulin, cells cannot absorb glucose, which they need to produce energy and thrive. In type 1 diabetes, an autoimmune disease, a person cannot make insulin because the body has killed the specialized cells in the pancreas needed to make insulin. For this reason, those with type 1 diabetes have to inject themselves with insulin to make up for their lack of being able to produce the substance on their own. Type 2 diabetes is a disease characterized by chronic high blood sugar that causes dramatic surges in insulin so high that cells become desensitized to the hormone. Think of it as being in a room where the volume of the music is cranked up so high that you feel the need to cover your ears. That's essentially what the cells do in the presence of too much insulin: They shut down the receptors that normally bind insulin and transport it inside. So while a person with type 2 diabetes can produce insulin, his or her cells don't use it as well as they should (we call this insulin resistance) and sugar remains in the blood, where it doesn't belong. Unlike type 1 diabetes, which is triggered by a faulty immune system, type 2 diabetes is mostly caused by diet—too much sugar and processed carbohydrates that force more insulin to be pumped out by the pancreas. And what the science is now revealing is that Alzheimer's disease could be another potential side effect of a sugary Western diet.

People with type 2 diabetes may be at least twice as likely to develop Alzheimer's disease, and those with prediabetes or metabolic syndrome may have an increased risk for having predementia or mild cognitive impairment (MCI).[8] Not all studies confirm the connection, but the evidence is mounting, forcing scientists to think differently and see broader relationships when it comes to risk for brain disease. The path from a poor diet to Alzheimer's doesn't appear to have to go through type 2 diabetes. In other words, studies are now showing that people with high blood sugar have a higher rate of cognitive decline than those with normal blood sugar. This was true in one particularly alarming longitudinal study following more than five thousand people over ten years.[9] Their rate of cognitive decline, regardless of whether they were diabetic, hinged on blood sugar levels. The higher the blood sugar, the faster the decline.

At the root of type 3 diabetes is the phenomenon that neurons in the brain become unable to respond to insulin, which means they can no longer absorb glucose, ultimately leading to cell starvation and death as insulin signaling becomes disrupted. Some researchers believe insulin deficiency or resistance is central to the cognitive decline of Alzheimer's disease and could be implicated in the formation of those infamous plaques.

A 2017 study from Dr. Guojun Bu, a Mayo Clinic neuroscientist and professor of medicine, found more evidence for type 3 diabetes when he showed that the variant of the Alzheimer's gene known as APOE4 is responsible for interrupting how the brain processes insulin.[10] APOE4 is found in approximately 20 percent of the general population and more than half of Alzheimer's cases. In Dr. Bu's study, the mice with the APOE4 gene showed insulin impairment, particularly in old age.

Put all of this information together, and it lends credence to the link of genetics, poor diet, and risk for cognitive decline. I find it interesting that not only have we witnessed a parallel rise in the number of type 2 diabetes cases and the number of people who are considered obese, but we've also begun to document the same pattern among those with dementia: As the rate of type 2 diabetes increases, so does the rate of Alzheimer's disease. Remember this, because it will explain some of the strategies in the Keep Sharp plan later in the book.

I should also add something about weight here, because we all know that there's often a relationship between weight and risk for diabetes. If the risk for Alzheimer's disease goes up with metabolic disorders, then it makes sense that the risk also rises with unhealthy weight gain that has metabolic consequences. The science now speaks to this fact. Carrying extra weight around the abdomen has been shown to be particularly harmful to the brain. One study that garnered lots of media attention looked at over six thousand individuals aged forty to forty-five and measured the size of their bellies between 1964 and 1973.[11] A few decades later, they were evaluated to see who had developed dementia and how that related to their waist size at the start of the study. The correlation between risk of dementia and thicker midsections twenty-seven years earlier was remarkable: Those with the highest level of abdominal fat had an increased risk of dementia of almost three-fold in comparison to those with the lowest abdominal weight. There is plenty of evidence that managing your weight now will go a long way toward preventing brain decline later.

Toxic Substances

More research is needed to understand which chemicals can result in brain abnormalities. I am not talking about well-known neurotoxins that can adversely affect brain function such as lead, tetanus toxin (from a bacteria), and mercury. I am talking about the exposure to chemicals we inadvertently encounter in our daily lives that could be inflicting harm slowly over time—for example, certain pesticides, insecticides, substances in plastic, food additives, and chemicals in our general household goods. For a long time, aluminum was feared as a "cause" of Alzheimer's disease, motivating many people to throw out their aluminum pots and pans. Although aluminum's neurotoxicity is incontrovertible, it is more difficult to establish a direct association between aluminum and Alzheimer's disease. Today, the theory that aluminum causes dementia has been largely discredited, but there are many other neurotoxins of concern, which is why future research will likely provide some answers.

In summer 2019, I went to Jackson Hole, Wyoming, to spend time with Paul Alan Cox, an ethnobotanist who studies the way indigenous people interact with their environment, particularly plants. His work took him to Guam, where he studied the native Chamorro people, who were known to be one hundred times more likely to develop a complex of neurodegenerative diseases, including Alzheimer's, compared to the rest of the world. Puzzled, he began putting his skills to work and created a consortium of scientists from a wide range of disciplines to investigate. What they found may one day be relevant to everyone. Because of their diet, which includes the fox bat as a delicacy, the Chamorro have been inadvertently poisoning themselves with BMAA, a neurotoxin produced by blue-green algae (cyanobacteria). While the Chamorro take it in high doses, because it is concentrated in the fox bat, it turns out that we are all exposed to BMAA, which could be a significant risk factor for Alzheimer's. The BMAA neurotoxin causes the proteins, such as amyloid and tau, to misfold and clump together in plaques and tangles. Because of this, Cox believes, as increasingly do other scientists, that amyloid and tau aren't the cause of Alzheimer's disease but a consequence of it. It's a big idea, but even more important is Cox's team's ongoing investigation into how to treat Alzheimer's disease in a remarkably simple way.

By replacing one of the building blocks of these proteins with an amino acid known as L-serine, they have shown that the misfolding of amyloid and tau doesn't continue happening, effectively halting the progression of Alzheimer's disease. So far, Cox's team has demonstrated this only in fervet monkeys, but human trials are now underway at Dartmouth College in New Hampshire. Best part of all, L-serine is widely available (in supplement form, usually a capsule), seems to have hardly any side effects, and costs just a few dollars. Cox will be the first to tell you that it is not a cure, meaning it won't reverse cognitive decline that has already occurred. Remember, however, that Alzheimer's disease typically starts in the brain long before someone develops symptoms. If a simple treatment could be given during this early stage, it could prevent someone from developing the symptoms in the first place. It's exciting work and further deflates the beta-amyloid hypothesis, providing more evidence that amyloid plaques could be a symptom of the disease, not the source.

Infections

Can infections earlier in life set the stage for Alzheimer's disease decades later? We have known for some time that infections from various pathogens can have neurological effects, from Lyme disease caused by the bacterium *Borrelia burgdorferi* to herpes simplex virus, Zika, syphilis, rabies, and even gum disease.[12] A hypothesis is now developing among scientists that serious forms of neurodegenerative decline can stem from the body's reaction to these infections.[13] This remains a hotly debated topic because we don't know if the germs' presence is causing or accelerating the disease or is merely a consequence of it. But the theory is plausible enough to attract the attention of top scientists.

A provocative study from Harvard researchers led by the late Dr. Robert D. Moir in 2016 proposed that infections, including mild ones that barely produce symptoms, fire up the immune system in the brain and leave a debris trail that is the hallmark of Alzheimer's.[14] The theory: a virus, bacterium, or fungus sneaks past the blood-brain barrier (which does become leaky as we age) and triggers the brain's self-defense system. To combat the intruder, the brain makes beta-amyloid to act as a kind of sticky web to trap the invader. The beta-amyloid is in fact an antimicrobial peptide—basically, a protein that the immune system creates to physically trap a germ. So what's left is the webby plaque that we see in the brains of Alzheimer's disease.

More work is needed in this area because not everyone who has had a brain infection develops Alzheimer's and not everyone who gets dementia can attribute the condition solely to an infection. Some people's brains may just be better genetically equipped to clear out those balls of beta-amyloid after they have killed the microbes, and other people's brains may be more vulnerable. Dr. Rudolph Tanzi, who is director of the Genetics and Aging Research Unit at MassGeneral Institute for Neurodegenerative Disease, is now leading the Brain Microbiome Project to learn what bacteria the brain can harbor and how to decipher the friendly colonies from the potentially harmful ones. When I spoke with Dr. Tanzi, who is also credited with discovering the Alzheimer's genes in the 1980s and 1990s, he clarified the connection between certain infections and Alzheimer's disease. See below, "Dr. Rudy Tanzi's 'Alzheimer's in a Dish.'"

DR. RUDY TANZI'S "ALZHEIMER'S IN A DISH"

Since 2014, scientists have made great strides in understanding Alzheimer's pathology thanks to Rudy Tanzi's "Alzheimer's in a dish," the world's first petri dish model of the disease. He and his team took mini–human brain organoids—clumps of brain cells used to develop "mini-brains"—grew them in a petri dish, inserted the Alzheimer's genes, and then watched what happened. This is how he observed the interplay between the plaques and tangles and then what followed: neuroinflammation, then significant nerve cell death. His metaphor is frightening but makes the point: "Amyloid plaques are the match, tangles are the brushfires, and neuroinflammation is the forest fire," he told me. Tanzi believes the brain's immune system tries to extinguish the brushfires by sending a surge of inflammatory cells. This neuroinflammation then kills up to one hundred times more nerve cells, laying the ground for future dementia.

According to Dr. Tanzi, this sequence of events helps explain why clinical trials have failed: They try to hit amyloid way too late. The best way to prevent a forest fire is to blow out the match first. The key is to stop amyloid from developing in the first place and target people before symptoms even develop.

So what strikes the match? Dr. Tanzi's lab has found that amyloid forms rather instantly around viruses like the herpes virus or bacteria and fungi like yeast. "Within twenty-four hours, a plaque forms with the virus trapped inside it. These are called extracellular traps and they're part of our innate immune system. Antibodies take a while to kick in when we get an infection, but before that, our primitive immune system tries to help us." While the immune system helps protect us at the time of infection, though, it may also set the stage for Alzheimer's later in life.

This doesn't mean you need a germ present to make a plaque. Other "ingredients" can cause plaques to form as well, and genetics certainly plays a role in making some people more likely to form plaques. This also doesn't mean certain germs definitively cause Alzheimer's. But what's interesting to note is that as we get older, our viral and bacterial loads from a lifetime of exposures are much higher than they were when we were kids. Some germs, such as the herpes simplex virus 1 that causes cold sores, can reactivate later in life. And when that happens, amyloid instantly gets seeded in a way that resembles cloud seeding. A big mass forms around the virus and traps it to protect the nerve cells in the brain. From Tanzi's perspective, we all need a little bit of beta-amyloid protein to protect the brain, but there may be a point when that protection can also pose a problem. As to why some people live with lots of brain plaques but never develop dementia? Tanzi calls those "resilient brains," and we'll get to the secrets to that later in the book. The key is making sure the brain's immune system doesn't overreact with neuroinflammation. I will teach you strategies to help with that as well.

Head Trauma and Injury

Repeated blows to the head can do lasting damage. Dr. Gary Small, founding director of the UCLA Memory Clinic, professor of psychiatry, and the director of the UCLA Center on Aging, as well as an expert for the Global Council on Brain Health, was the doctor who diagnosed Tony Dorsett's CTE. Dr. Small's group's finding was among the first to link multiple concussions to damaging tau buildup. Dorsett had suffered from depression and memory loss for many years and went to UCLA for answers. He wanted to know whether there could be a connection between all the concussions he endured in the 1970s and 1980s playing football and the debilitating symptoms he suffered later in life. Since Dorsett's diagnosis, scores of other former football players have been diagnosed with CTE, and lawsuits have been filed against the National Football League. Gary Small has been a pioneer in brain medicine for decades, and I've had the opportunity to consult with him on his research and findings. You'll read more about his top strategies to keep sharp in part 2.

Immune System Challenges and Chronic Inflammation

I've already covered the immune system's potential role in neurodegeneration and the downstream effects of inflammation. It is worth highlighting a few more specifics because chronic inflammation associated with aging ("inflamm-aging") is at the center of virtually all degenerative conditions, from those that increase one's risk for dementia, such as diabetes and vascular diseases, to those that are directly brain related, such as depression and Alzheimer's disease. For decades scientists have debated the role of inflammation in a diseased brain, but now a burst of new research suggests that inflammation not only adds to disease processes in the brain that cause decline, but it also ignites those processes in the first place. One new study published in 2019 out of Johns Hopkins showed that chronic inflammation at midlife is linked to later cognitive decline and Alzheimer's disease.[15]

To be sure, inflammation is the body's defense system for taking care of potential insults and injury, but when that system is constantly deploying chemical substances and keying up the immune system, it becomes problematic. Although studies in the past showed that people who have taken common anti-inflammatory medications such as Advil (ibuprofen) and Aleve (naproxen) for two or more years may have a *reduced* risk for Alzheimer's and Parkinson's disease, subsequent clinical trials have failed to show that these anti-inflammatory drugs can significantly lessen or totally prevent Alzheimer's, and taking them comes with their own side effects and other risks.[16] At the same time, other studies have shown heightened levels of cytokines in the brains of individuals suffering from these and other degenerative brain disorders. Cytokines are the substances secreted by cells in the body that act like traffic signals for the inflammatory process, among other things. What this means is that chronic inflammation likely plays a big role in brain decline. Today, new imaging technology is finally allowing us to see cells actively involved in producing inflammatory cytokines in the brains of Alzheimer's patients.

Inflammation in the brain can also be directly related to those amyloid plaques and tau tangles, which again shows how interconnected and interrelated some of these "causes" for Alzheimer's can be. Specialized "housekeeping" or "support staff" cells in the brain called *microglia*, or simply glia or glial cells as I previously defined, sometimes recognize these proteins as foreign debris and release inflammatory molecules to get rid of them. Glial cells are the brain's unique immune cells and are related to types of white blood cells called macrophages. The resulting inflammation from the glial cells' actions further impairs the working of neurons, thereby worsening the disease process. But, once again, the exact cause-and-effect mechanism remains a mystery. We can't say for sure that inflammation directly causes Alzheimer's disease, though it's likely to be a big part of the whole picture.

TYPES OF COGNITIVE DEFICITS

Just as different types of cognitive deficits can be experienced, there is not a well-defined road to full-blown Alzheimer's from a normal aging brain. Let's take a look at the terms often used to differentiate certain conditions from others. Alzheimer's disease is one type of dementia, and the individual experience with it can vary greatly from person to person. According to the Alzheimer's Association, up to 40 percent of dementias are caused by conditions other than Alzheimer's.[17]

Normal Aging

Your brain, like the rest of your body, changes as you grow older. While there is normal age-related tissue loss and degeneration of the synapses, here's a new finding we all should rejoice over. In 2018, researchers from Columbia University showed for the first time that healthy older folks can generate just as many new brain cells as younger people.[18] The researchers found that the ability to make new neurons from precursor cells in the hippocampus, the brain's memory center, does not hinge solely on age. Although older folks have less vascularization (fewer and less robust blood vessels) and perhaps less ability of new neurons to make connections, they don't necessarily lose their ability to grow new brain cells. The key word here, though, is *healthy*—as in *healthy individuals*. It should be clear by now that to maintain neurogenesis, vascularization, and make new neural connections, you have to stay healthy overall. This is another reason the connection between the mind and the body is so strong.

It's important to bear in mind that the brain begins to age in one's mid-twenties and can structurally begin to deteriorate as early as age thirty. After age forty, the hippocampus shrinks by about 0.5 percent per year. This shrinkage, however, is quite variable among individuals and heavily depends on lifestyle choices, environmental factors, genetic predisposition, and medical conditions. These factors have an impact on the hippocampus more than any other part of the brain. Dozens of neuroscience research studies have shown that the hippocampus is fragile and shrinks more than any other brain area—with whatever insults happen to the brain. For example, traumatic brain injury, diabetes, or vitamin B12 deficiency results in atrophy in the hippocampus more so than other brain regions.

We all experience a breakdown of the assembly process of memory described earlier, and that breakdown can begin in a subtle way when we are young, worsening as we reach our fifties and beyond. I have seen the physical changes of an aged brain upon autopsy. The brain has shrunk, the folds are more prominent, and the blood vessels are hardened and less robust. Under the microscope, you may also see evidence of neuronal cell death and even changes in the synapses. Still, none of this necessarily correlates with outward signs of cognitive decline while the individual was alive. The point is that there has been a conceptual shift away from the view of aging as a disease, even if aging is a risk factor for certain diseases. In other words, aging does not mean there will be inevitable cognitive decline. Any cognitive decline, be it "normal" or abnormal, is more than just a factor of age and brain degeneration.

Mild Cognitive Impairment

MCI is often the beginning stage of dementia, but not everyone with MCI will go on to develop a more severe form or Alzheimer's disease. They are simply at an increased risk. MCI causes a slight, often unnoticeable, decline in memory function. An example is a seventy-five-year-old person who repeats the same question five or six times in an hour but can still drive and manage daily routines. Unlike other types of cognitive impairment that affect speech and bodily control, with MCI only memory is impacted. It's important to treat signs and symptoms as early as possible. Nearly 10 to 20 percent of people sixty-five and older are estimated to have MCI.[19]

Dementia

The term *dementia* is a general one used to describe various symptoms and severities of cognitive decline, starting with mild cognitive impairment and moving to severe dementia. In other words, dementia is not a single disease in itself; it encompasses several underlying diseases and brain disorders that impair memory, communication, and thinking. There are several types of dementia.

Vascular Dementia. This type of dementia is caused by an impaired blood supply to the brain and may be brought on by a blood vessel blockage or damage leading to strokes or bleeding in the brain. Sometimes a person can show symptoms of both vascular dementia and Alzheimer's disease at the same time. The location and amount of the brain damage determines whether dementia will result and how the individual's thinking and physical functioning will be affected. It used to be that evidence of vascular dementia was used to exclude a diagnosis of Alzheimer's (and vice versa). That practice is no longer used because the brain changes of Alzheimer's and of vascular dementia commonly coexist. Only about 10 percent of brains from individuals with dementia show evidence of vascular dementia alone, and about half of all people with Alzheimer's have signs of silent strokes.[20]

Dementia with Lewy Bodies (DLB). This condition affects about one in every five patients with dementia. Proteins, called alpha-synuclein or Lewy bodies, will build up in certain parts of the brain responsible for cognition, movement, and overall behavior. As a result, patients have memory problems and symptoms similar to Parkinson's. Visual hallucinations often occur early and can be an important clue for the diagnosis.

Frontotemporal Lobar Dementia (FTLD). Also known as Pick's disease, FTLD is a group of disorders triggered by gradual nerve cell loss in the brain's frontal and temporal lobes, resulting in changes in behavior (e.g., socially inappropriate responses, loss of empathy, lack of inhibition, poor judgment), difficulty speaking, and memory problems—though memory is usually spared

Memory Loss, MCI, and Dementia (Alzheimer's Disease)

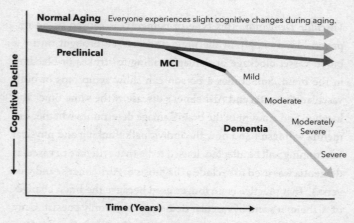

The Road to Severe Dementia

in the early stages of this disease. The changes in personality and behavior are often the first signs. About 60 percent of people with FTLD are forty-five to sixty years old, but FTLD accounts for only 10 percent of dementia cases.[21]

Alzheimer's Disease. This is the most common form of dementia. It is a progressive disease with symptoms that typically develop gradually before they intensify and become severe. In its late stages, the disease can make it difficult for a person to handle daily tasks, think clearly, control bodily movements, and live independently. Alzheimer's disease accounts for 60 to 80 percent of dementia cases, affects one in nine Americans age sixty-five and older, and is the sixth leading cause of death in the United States. Nearly 6 million people are living with this disease. When someone shows signs of having both Alzheimer's and other dementias, it's called mixed dementia.[22]

NORMAL VERSUS NOT NORMAL

You forgot what day of the week it was when you woke up this morning. Normal or a sign of something serious? You can't remember the phone number you had twenty years ago or the name of your high school track coach. Typical? One of the first questions people ask themselves upon forgetting something seemingly basic or being unable to recall a long-lost classmate's first name at a reunion is: Is this normal or the first stages of cognitive decline? Mary A. Fischer explains six types of normal memory lapses for AARP that are not a cause for worry. Whew![23]

Absentmindedness. Where did you put your keys down? Or what was the reason to walk into the dining room? We all experience this on occasion, and we can blame a general lack of attention or focus. It's normal to forget directions to somewhere you haven't visited in a while. But if you finish grocery shopping at your usual store and then can't find your way home, that could be a problem that is beyond absentmindedness. In their seminal work, *The Memory Book*, Harry Lorayne and Jerry Lucas eloquently describe the important process of what they call establishing "original awareness."[24] They use the term simply to refer to the "first time"—as in the first time you see or do something that you want to remember. When you set your keys down on a table, you need to have an original awareness when you initially set them down in order to remember that you put them there. You need to actively observe what you're doing. In fact, observation is essential to original awareness, and it's not the same as just "seeing." There is a difference between what the eyes "see" and what the mind "observes." If your mind is "absent" when you perform an action, there can be no observation; more important, there can be no awareness of the action (learning) and subsequent creation of the memory.

Blocking. This is the classic but frustrating experience of not being able to recall something from memory that you feel is right there. You know what you're trying to say, but it's hidden. Blocking usually results from several similar memories creating a disruption. Multiple studies have shown that older participants often activate more areas of the brain to perform a memory task than the study's young subjects.[25] Think of it as if your memory retrieval button gets jammed from time to time.

Scrambling. If you've ever gotten the details wrong but can accurately remember most of an event or other chunks of information, you're scrambling—confusing those little details. For example, a good friend tells you that she is taking a writing class to finish her novel. Later, you correctly recall this bit of information but think she told you in person when it was during a phone conversation. A glitch in the hippocampus is likely to blame for this. It has incorrectly recorded the time and place of the facts.

Fading away. The brain continually cleans out older memories to make room for new ones. Memories that are not recalled often can begin to fade away because those memories are not being reinforced. Which is why it's relatively easier to remember the details of what you did more recently than what happened many years ago. This basic use-it-or-lose-it characteristic of memory is called transience, and it's normal at all ages.

Struggling for retrieval. This one is similar to absentmindedness. You just met someone for the first time, and seconds later, you can't remember her name. Or you saw a great movie, but when you tell a friend about it the next day, you've completely forgotten the title or the starring actor's name. Aging changes the strengths of the connections between neurons in the brain, and new information can delete other items from short-term memory unless it is repeated again and again. This is why paying special attention to learning someone's name on the spot and associating that name with something particular or familiar will help you avoid this glitch.

Muddled multitasking: At some point, the number of things you can do effectively at one time diminishes. Maybe you can't type an email and watch TV at the same time. Studies show that the older we get, the more effort it takes the brain to maintain focus, and it takes longer to get back to an original task after an interruption. We'll see in chapter 6 how ending attempts to multitask can actually be a good thing for the brain.

RETHINKING COGNITIVE DECLINE

Is Alzheimer's disease overdiagnosed? It's a provocative question, and one that can lead to a surprisingly uplifting idea. Because there is no definitive way to diagnose Alzheimer's as you can diabetes or heart disease, it is possible to saddle people with this label too quickly. For some people, the reality is they can reverse their cognitive decline because they never had Alzheimer's in the first place. It's a point Dr. Majid Fotuhi raised with me in a lively discussion, and it's worth considering his perspective.

Dr. Fotuhi is a neurologist and neuroscientist with more than twenty-five years of research and clinical experience at Johns Hopkins and Harvard Medical School in the field of memory, aging, and brain rehabilitation. Today he treats patients with a wide range of complicated neurological issues, from cognitive impairments to postconcussion syndrome, vertigo, chronic migraine, and attention deficit disorders. He reports that he has achieved remarkable results when he puts patients through his multidisciplinary protocols tailored to each unique individual. For him, a comprehensive brain fitness program focuses on lifestyle strategies to modify risk factors such as vascular disorders, vitamin deficiencies, obesity, diabetes, depression, anxiety, sleep apnea, and sedentary behavior. In his own research, he has documented vast improvements in patients who at one time felt hopeless about the future of their brains. But he proved them wrong by letting the results speak for themselves. He has even documented substantial growth in volume of that all-important memory center of the brain, the hippocampus, within weeks of an interventional program.

The suggestions I make for you to do at home will no doubt echo some of the protocols he offers to C-suite executives around the country who have access to his exclusive care. "I want to change the conversation," says Dr. Fotuhi. By focusing on brain growth and repair rather than telling people they have a fatal condition, he hopes more people will be inspired by the possibility of building bigger, better brains today. He goes so far as to suggest we do away with the doomsday term *Alzheimer's disease* and instead establish new terminology that simply uses words like *mild, moderate,* and *severe cognitive impairment.* Like many of the other researchers I spoke to for this book, Dr. Fotuhi is critical of the amyloid cascade hypothesis as the basis for all patients with Alzheimer's. In his

2009 *Nature Review* article, he provided an alternative theory, the dynamic polygon hypothesis.[26]

This is how he explains it: "Multiple risk factors—and protective factors—interact with each other to help us either stay sharp with aging or decline rapidly. I continue to believe that it is naive to target amyloid as the sole culprit for a decline that happens to most people in late life—with variable speed and with many different clinical manifestations. Amyloid is the sole culprit only in patients with early-onset Alzheimer's disease—which is quite different than late-life 'Alzheimer's disease.'" Remember this point. For many patients diagnosed with cognitive decline, the reality is they may have neither amyloid nor Alzheimer's.

FOCUS ON YOUR BRAIN, AND EVERYTHING ELSE WILL FOLLOW

When I interviewed top experts in brain health from a wide variety of professionals and pioneers in the field, one individual's statement stood out from the rest. It came from Dr. Dan Johnston, a former lieutenant colonel in the U.S. Army who has served as a physician and researcher in the army from the Pentagon to Iraq and recently cofounded BrainSpan, a company and laboratory that develops products and programs to help people measure, track, and improve brain function. As a health care provider, his company delivers its products mostly through physicians.

To say Johnston's goal is to optimize brain health and performance is an understatement. He aims to shift the way we think about health by "starting at the top," as he puts it. When it comes to health, many of us immediately turn to things like weight, cholesterol levels, risk for cancer, blood sugar levels, and heart health, and we forget about the brain. Those other things are seemingly easier to grasp because the brain is encased in bone and has a mystical quality. The medical establishment has typically interfaced with the brain only when it is diseased or damaged. But here's the key point: when you put your brain first, everything else health-wise falls into place. The brain is ground zero. Don't forget that it is what makes you. Your heart ticks, yes, but it's your brain that ultimately makes you tick and determines your quality of life. Without a healthy brain, you cannot even make healthy decisions. And with a healthy brain comes not only a healthy body, weight, heart, and so on, but also a stronger sense of confidence, a more solid financial future thanks to smart decisions, better relationships, more love in your life, and heightened overall happiness.

The upcoming chapters put the brain first. If you're worried about something else—maybe those extra twenty pounds, the general aches and pains, the insomnia and chronic headaches—challenge yourself to make brain health a priority and watch what happens.

CHAPTER 3

12 Destructive Myths and the 5 Pillars That Will Build You

Overall, the human brain is the most complex object
known in the universe—known, that is, to itself.

EDWARD O. WILSON

As a neurosurgeon, I get to live a life that has a clear sense of purpose. Patients come into the hospital under dire circumstances, and they put all their faith in me. It is an awesome responsibility. After practicing for nearly twenty years, I am still thrilled to speak to a family after a successful operation—whether it's removing a tumor, clearing pooled blood after trauma, or repairing a fractured spine. But I also make my living by taking my expertise on the road and putting on my journalist's hat to report from the front lines of a newsworthy event. And when my worlds of medicine and media collide, the result can be spectacular.

In spring 2003, I was in Iraq for several weeks with a group of doctors known as the Devil Docs—navy doctors who helped support the marines. We had spent countless days together traveling through the desert, caring for patients terribly wounded, and getting to know each other well under incredibly unique and challenging circumstances. One day, some of the Devil Docs ran over to me and asked if I would literally take off my journalist's cap and put on a surgeon's cap. A young lieutenant had been shot in the back of the head and was thought to have been mortally wounded, but as they brought his body to the Devil Docs' camp, his pulse returned. He was alive but in need of quick surgery. Time was of the essence, and I was the only neurosurgeon in the area, so they wanted me to help. I rushed him to the makeshift operating room and realized that he needed a craniectomy—removal of a portion of his skull to relieve pressure on his brain and drain the blood collection. Without proper tools in the dusty, desert tent, I took the bit from a Black & Decker drill and sterilized it. I placed a sterile glove on the drill itself, and then used it to open his skull and provide room for his swollen brain. After that, I dissected through the outer layers of the brain, found the blood clot and shrapnel, and carefully removed it. I still needed to cover his brain in some sterile fashion, otherwise he would have been at risk of meningitis, possibly encephalitis, and most likely he would not have survived. So I fileted open an IV bag and used the inside of it to re-create the outer layer of his brain, because it was the only thing truly sterile inside that dusty tent.

After that, I wrapped his head, and he was evacuated by a Black Hawk helicopter to Kuwait. I wasn't sure I would ever see him again or even if he would survive. A couple of months later, a doctor called me from San Diego to give me an update on the young man, Jesus Vidana. He was alive and doing well, he told me. I visited with him a short time later, and I invited him to be my guest when I gave the commencement address at the University of Southern California Medical School. He received a long standing ovation, and it still gives me chills to think of his beaming face, handsome and healthy. His survival under those conditions with the gravity of his wounds was one of the most electrifying experiences of my life. I like to joke that operating on Jesus in the middle of the desert is something I will never forget!

The reason I share this story is that it highlights what's possible when a brain—against all odds—survives a trauma. It's more resilient and recoverable than you think. And you can take action to reverse the course of a brain's inevitable demise under even bad circumstances. The example is extreme, but remember it as you move forward and learn the ways in which you can shift your own circumstances to decrease your chances of ever facing a brain-related condition or, worse, dying from a brain-related ailment.

THE DIRTY DOZEN

You've already gained a lot of wisdom about the brain in previous chapters. But I bet you'd still get a few facts wrong when it comes to answering questions about what it can do and how it changes throughout your lifetime. Remember that I want you to know the why and the how when it comes to brain health. Let's add to your working knowledge by debunking the twelve most pervasive myths about the aging brain. This will ultimately prepare you to embrace what you can do to de-age your brain and add more years to its health. I call these myths the Dirty Dozen.

Myth #1: The Brain Remains a Complete Mystery

I have a love-hate relationship with this myth. I hate it because it's not true, but I love it because it allows me to correct people's misconceptions and give them hope. While there is still a lot to learn, researchers have recently made great strides in understanding the brain. We know more about the connections between different parts of the brain and their relevance to how we think, move, and feel. We are better able to anatomically identify the areas of the brain responsible for depression, obsessive compulsive disorder, and addiction. And we can better rehabilitate the brain after injury or stroke. The field of neuroscience is almost constantly brimming with new and exciting breakthroughs, many of which I highlight in part 2.

Myth #2: Older People Are Doomed to Forget Things

There is a kernel of truth to this myth; some cognitive skills do decline as you age, especially if you don't employ strategies to pay closer attention and help you remember. But whereas you may have been quicker at picking up a new language or memorizing a list of random words when you were a youngster, you're more likely to be superior with vocabulary and a good judge of character when you're an older adult. You'd score higher on tests of social communication and diplomacy, such as how to settle an argument or deal with a conflict. The other good news about aging is that we tend to improve over time at controlling our own emotions, weathering stress, and finding meaning in our lives.

Myth #3: Dementia Is an Inevitable Consequence to Old Age

You should be able to dispel this myth on your own by now. Dementia is not a normal part of aging. Typical age-related changes in the brain are not the same as changes that are caused by disease. The former can be slowed down and the latter can be avoided.

Myth #4: Older People Can't Learn New Things

Learning can happen at any age, especially when you get involved with cognitively stimulating activities like meeting new people or trying new hobbies. The combination of memory being dynamic and the possibility of growing new neurons (neurogenesis) means we can continue to change our brain's information, capacity, and learning strengths. While mastering some new skills, such as a second or third language, may take an older person longer, that doesn't mean one cannot achieve the feat. Never say "never." Even people diagnosed with cognitive decline, Alzheimer's disease included, can continue to learn new things.

Myth #5: You Must Master One Language before Learning Another

Young children who learn English and a different language at the same time do not confuse the two, and even though they may take longer to master both at the same time, that does not mean it's a bad idea. Different areas of the brain do not go to battle so there's no interference. Much to the contrary, kids who are bilingual have a better generalized knowledge of language structure as a whole. One of the reasons children seem to learn a new language more easily than adults is that they are less self-conscious.

Myth #6: A Person Who Has Memory Training Never Forgets

In part 2, I present a gallery of memory training ideas to build those skills. One of them is "use it or lose it." It applies to memory training in the same way it applies to maintaining the strength of a muscle or your overall physical health. This will be an ongoing practice that you'll need to maintain, as with other long-term strategies.

Myth #7: We Use Only 10 Percent of Our Brains

Who hasn't heard of this myth? It's been around for a long time, suggesting that we have huge reserves of untapped mental powers. But do we really waste 90 percent of our brains? Absolutely not. That would be ridiculous just from an evolutionary standpoint. Brains are demanding organs; it takes a lot of energy to build them during development and maintain them as adults. Evolutionarily, it would make no sense to carry around surplus brain tissue (and let's apply some logic: if the 10 percent idea were true, it would certainly make brain damage a lot less worrisome). Experiments using positron-emission tomography (PET) or functional magnetic resonance imaging (fMRI) scans show that much of the brain is engaged during even simple tasks, and injury to the small sections of brain called "eloquent areas" can have profound consequences for language, movement, emotion, or sensory perception.

Remember that autopsy studies show that many people had physical signs of Alzheimer's disease (such as amyloid plaques among neurons) in their brains even though they had no symptoms. Perhaps we can indeed lose some brain tissue and still fully function. There is something to be said, however, about choosing to exercise our minds at 100 percent capacity. People score higher on IQ tests if they're highly motivated, and that's not surprising. I think of the brain like a town. The important structures such as the homes and shops are in nearly constant use, and they probably represent 10 to 20 percent of our brains. The rest, however, are the roads that connect all these shops and homes. Without the roads, information could not get where it needs to go. So while the roads are not in constant use, they are necessary.

Myth #8: Male and Female Brains Differ in Ways That Dictate
    Learning Abilities and Intelligence

Urban legend has it that men are biologically better suited for math and science, whereas women are better suited for empathy and intuition. Some of the worst-designed, least reproducible, and most biased research in the history of science claims to provide biological explanations for differences between the sexes. Granted, differences do exist in the brains of males and females that result in variations in brain function, but not to the extent that one is better "equipped" than the other. Scientists continue to study the brain to understand and learn more about any important differences between the brains of men and women—research is still emerging in the field of neuroscience. Another way to think about it more broadly: Each of us may be wired in our own unique way, though with a healthy brain we each have the capacity to learn, remember, and make sense of the complex world around us.

One item to note, however, is that Alzheimer's strikes a disproportionate number of women compared to men.[1] Two-thirds of Americans with Alzheimer's disease are women, and we don't have an understanding yet why this is the case or what causes women to be at higher risk. It's not just because they are more likely to live longer. Something about their physiology could be part of the reason. How many times a woman is pregnant over her lifetime is among the provocative theories being studied.[2] Pregnancy entails many biological events from hormonal changes to immune function shifts that could ultimately lead to protection against developing dementia later in life. We don't have the answers yet, although hormone therapy continues to be discussed as a tool. It has been shown to be harmful for cognition under certain circumstances but potentially beneficial under other circumstances, depending on when you start taking it (in one's early fifties or between sixty-five and seventy-nine, respectively). What's becoming clear is that a personalized approach needs to be considered. Different women respond differently to hormonal therapy depending on their individual risk factors, such as being diabetic or carrying a gene related with Alzheimer's.

Women do have an advantage over men in their verbal abilities, and that could be a factor in identifying any cognitive problems. Studies show women score better on standard tests used to diagnose the early stages of dementia, even when brain scans suggest they are at the same stage of the disease as men.[3] Put simply, women can hide symptoms of Alzheimer's with their superior verbal skills, which results in their not getting diagnosed early enough. At later stages of cognitive impairment, this advantage disappears. Such a gender-based difference may be the reason that women seem to decline more rapidly after being diagnosed—they are further along the disease's trajectory than the earlier test would have indicated. Calls for gender-based cutoff points for these tests are now part of the conversation in research and clinical settings. (I discuss this in more detail with Maria Shriver in chapter 11.)

Myth #9: A Crossword Puzzle a Day Can Keep the Brain Doctor Away

Another urban legend is that doing crossword puzzles will keep your brain young. Unfortunately, crossword puzzles flex only a portion of your brain, mostly its word finding ability (also called fluency). So while they might help you excel at that, they won't necessarily keep your brain sharp in any general, overall sense. That said, there is value in doing word and number puzzles, including games like Sudoku. In 2019, a follow-up study by the University of Exeter Medical School and King's College London confirmed earlier results that showed the more often participants engaged with puzzles, the better they performed on tasks assessing attention, reasoning, and memory.[4] Their results came from analyzing data from more than nineteen thousand healthy people aged fifty and over who are enrolled in the large PROTECT Study, which spans twenty-five years and follows participants annually to explore how the brain ages and what might influence the risk of dementia later in life. The researchers were quick to point out that the results don't directly and definitively mean that doing crosswords improves brain function or gives you a sharper brain. What's known is that keeping an active mind can help to reduce decline in thinking skills, and for some people, doing crossword puzzles is a way to do that. For others, it may not be.

Myth #10: You Are Dominated by Either Your "Right" or "Left" Brain

Contrary to what you might have been taught in the past, your brain's "two sides"—right and left—are intricately codependent. You may have heard that you can be "right-brained" or "left-brained"—and that those who favor the right are more creative or artistic and those who favor the left are more technical and logical. The left brain/right brain notion originated from the realization that many people express and receive language more in the left hemisphere and spatial abilities and emotional expression more in the right. Psychologists have used the idea to distinguish between different personality types. But brain scanning technology has revealed that the brain's two hemispheres most often work together intricately. For example, language processing, once thought to be the domain of the left hemisphere only, is now understood to take place in both hemispheres. The left side handles grammar and pronunciation, while the right processes intonation, and the brain recruits both left and right sides for both reading and math.

Myth #11: You Have Only Five Senses

You can likely name all five senses: sight (ophthalmoception), smell (olfacoception), taste (gustaoception), touch (tactioception), and hearing (audioception). But there are others with the "cept" ending, which is Latin for take or receive. The other six senses are also processed in the brain and give us more data about the outside world:

- Proprioception: A sense of where your body parts are and what they're doing.
- Equilibrioception: A sense of balance, otherwise known as your internal GPS. This tells you if you're sitting, standing, or lying down. It's located in the inner ear (which is why problems in your inner ear can cause vertigo).
- Nociception: A sense of pain.
- Thermo(re)ception: A sense of temperature.
- Chronoception: A sense of the passage of time.
- Interoception: A sense of your internal needs, like hunger, thirst, needing to use the bathroom.

Myth #12: You're Born with All the Brain Cells You'll Ever Have, Your Brain Is Hardwired, and Brain Damage Is Always Permanent

If you have thought the head of a newborn looks proportionally bigger compared to its body size than the head of a grown adult, you're right. Due to the imbalance between brain and body development during pregnancy, babies' brains are proportionately much larger than those of adults relative to their body size. A newborn's brain triples in size in the first year of life; after that, the rate of physical growth slows as we learn and pack more into our roughly 3.3-pound brains. What continues to develop, allowing this tremendous ability to process more and more information, is the complexity of the networks of neurons as they go through a process of pruning, whereby certain synapses that are not being used are trimmed to make room for new ones. This helps explain why brain size is not necessarily directly correlated with intelligence. As the brain reaches half of its adult size by nine months and nearly three-quarters by two years of age, a baby's head must be large and grow rapidly to accommodate the rest of the body's growth. On average, the brain reaches its maximum size in girls at about eleven and a half years of age and fourteen and a half on average for boys—but again it will not be fully mature in terms of its internal development and executive functioning until about twenty-five years of age.

You know that as an adult, adding more information to your brain doesn't increase the size of it (and imagine what people would look like if brain size increased with learning new information). But what does grow larger is the number of neurons—nerve cells—and the complexity of their network through ongoing and active pruning and "growth." While genes likely play a role in the decline of synapses, among the most astonishing recent research is that which has highlighted the power of experience—how one's environment can profoundly influence the pruning process. It's the old nature-versus-nurture phenomenon in action. Synapses that are "exercised" by experience become stronger, while others get weaker and are eventually trimmed away.

As I've already noted, we used to believe that we were born with a finite number of neurons for life. When we damaged any of them, we couldn't replace them. Similarly, many scientists thought that the brain was unalterable: Once broken, it could not be fixed. Now we know differently. The brain remains plastic throughout life and can rewire itself in response to your experiences. It can also generate new brain cells under the right circumstances. Take, for example, what blind people experience, as parts of their brain that normally process sight may instead be devoted to exceptional hearing. Someone practicing a new skill, like learning to play the violin, "rewires" parts of the brain that are responsible for fine motor control. People who've suffered brain injuries can recruit other parts of their brain to compensate for the lost or damaged tissue. Intelligence is not fixed either.

Neurogenesis has long been proven in various other animals, but it wasn't until the 1990s that researchers began focusing exclusively on trying to demonstrate the birth of new brain cells in humans. Finally, in 1998, Swedish neurologist Peter Eriksson was among the first to publish a now widely cited report documenting that within our brains—in the hippocampus—there's a reservoir of neural stem cells that are continually replenished and can differentiate into brain neurons.[5] We all experience development, at least in certain areas of our brains, throughout our lives. We are also equipped with the technology to rewire and physically reshape our brains. This has led to the burgeoning new field of neuroplasticity—the ability of the brain to form and reorganize synaptic connections. The brain's plasticity was first documented more than 100 years ago in William James's 1890 book *The Principles of Psychology*, in which the Harvard University psychologist writes: "Organic matter, especially nervous tissue, seems endowed with a very extraordinary degree of plasticity," but only in my lifetime have we begun to measure and visualize this phenomenon with technology. And with tools like fMRI, we can see the brain changes in response to certain stimulation. We can also see parts of the brain that are not in use being pruned away. The brain constantly and dynamically shapes and reshapes itself in response to experiences, learning, or even an injury. What's more, what you choose to focus your attention on rewires the brain from a structural and functional perspective.

The fact neurogenesis occurs in us throughout our lifetimes, coupled with the bonus fact we can change its circuitry through neuroplasticity, has stirred a revolution in neuroscience and our thinking about the brain. This new knowledge also has instilled hope in those searching for clues to slowing down, reversing, or even stopping and curing progressive brain disease. If we can regenerate brain cells and reshape connections, imagine what that can do for the study of neurodegenerative disorders. My educated guess is that novel treatments are on the way. Some have already transformed the lives of people who have suffered from serious brain injuries or disease. Look no further than Sharon Begley's *Train Your Mind, Change Your Brain* to read about real-life stories that prove just how pliable our brains are.[6] Dr. Norman Doidge tells similar stories in his books that chronicle how the brain changes itself. If people who have suffered a devastating stroke can learn to speak again and those born with partial brains or who lose significant brain tissue to disease or surgical removal can propel their brains' rewiring to work as a whole, think of the possibilities for those of us who just hope to preserve our mental faculties as we age. Even people who've had an entire hemisphere removed in childhood to treat rare neurological conditions such as intractable epilepsy or brain cancer can go on to function in adulthood. Their brains reorganize and various networks pick up the slack.

If you're wondering just how the brain "grows" new neurons, it's largely through the help of the protein brain-derived neurotrophic factor (BDNF), which is encoded in a gene located on chromosome 11. Dr. John Ratey, a neuropsychiatrist at Harvard who has written extensively on the connection between physical fitness and brain health, calls BDNF "Miracle-Gro for the brain."[7] In addition to nurturing neurogenesis, BDNF also helps protect existing neurons and encourage synapse formation—the connection of one neuron to another. Interestingly, studies have demonstrated decreased levels of BDNF in Alzheimer's patients. It's no surprise, then, that scientists are looking for ways to increase BDNF in the brain through basic lifestyle habits. Among the things included on their list of upcoming strategies are exercise, restorative sleep, stress reduction, and healthy exposure to sunlight.

It's important to note that brain plasticity is a two-way street. In other words, it's almost as easy to drive changes that impair memory and physical and mental abilities as it is to improve these things. I love how Dr. Michael Merzenich, a leading pioneer in brain plasticity research and professor emeritus at the University of California at San Francisco, puts it: "Older people are absolute masters at encouraging plastic brain change in the wrong direction."[8] You can change your brain for the better or worse through behaviors and even ways of thinking. Bad habits have neural maps that reinforce those bad habits. *Negative plasticity*, for example, causes changes in neural connections that can be harmful. Negative thoughts and constant worrying can promote changes in the brain that are associated with depression and anxiety. Repeated mental states, where you focus your attention, what you experience, and how you respond to situations indeed become neural traits. One of Dr. Merzenich's often-cited quotes is the following: "The patterns of activity of neurons in sensory areas can be altered by patterns of attention. Experience coupled with attention leads to physical changes in the structure and future functioning of the nervous system. This leaves us with a clear physiological fact . . . moment by moment we choose and sculpt how our ever-changing minds will work. We choose who we will be in the next moment in a very real sense, and these choices are left embossed in physical form in our material selves."[9]

Secrets of SuperAgers

While it would be great to have the brain of a SuperAger, someone who has an uncanny ability to maintain a youthful brain well into old age, most of us didn't win the genetic lottery. A small, elite group of people age eighty and older have memories that are as sharp as those of people twenty to thirty years younger; they show no age-related shrinkage in the size of the brain networks correlated with memory ability.[10] And their outer cortexes, where memory, attention, and other thinking abilities take place, are remarkably thick—similar to people in their fifties. Scientists are trying to uncover their secrets to make us all SuperAgers, and they are learning it may not be totally genetically driven. What the science is increasingly showing is that we can have a huge impact in our brain's fate with simple lifestyle choices. SuperAgers often don't act like old folks. They too keep sharp with good habits.

HOW TO KEEP A SHARP MIND

Part 2 of this book covers the five pillars of brain health. It's how you'll continue to move your mind in the right direction. You'll gain a great understanding about the science behind these pillars and how to apply them easily in your own life. And for those willing to take the challenge, I will show you how to take each recommendation up a notch or two to truly optimize your brain. Not all of the strategies I suggest will be for everyone, but I trust that I've got something for everyone. I will even give you a program for those who need specific directions. (I can already hear some of you pleading: *Please tell me exactly what to do and not do.*) Finally, I'll give extra tips for those seeking to increase productivity, make the most of their time (like how to find a full *extra* hour in your day), and break bad habits while becoming the architect of good ones. At the core of the lesson is the goal: shape a better life through a sharper brain.

As a primer, here are the five pillars of brain health: Move, Discover, Relax, Nourish, Connect. These five pillars were first described by AARP based on the existing scientific evidence that demonstrated these actions are fundamental to promoting good cognitive function across the life-span. I recommend them to you to keep your mind sharp no matter your age. Here's what they mean in no particular order:

> *Move.* This should not come as a surprise. Exercise, both aerobic and nonaerobic (strength training), is not only good for the body; it's even better for the brain. Every day before sitting down to write this book, I made sure to do something physical. A bike ride, push-ups, a swim, or a run. If my writing ever starts to drag or not connect the way I want, I exercise my body as a way of stimulating my mind. Physical exertion, in fact, has thus far been the only thing we've scientifically documented to improve brain health and function. While we can record associations between, say, eating a healthy diet and having a healthier brain, the connec-

tion between physical fitness and brain fitness is clear, direct, and powerful. Movement can increase your brainpower by helping to increase, repair, and maintain brain cells, and it makes you more productive and more alert throughout the day. There is a nearly immediate measurable cause and effect going on that you'll soon learn about, and it's stunning. I have always followed the advice of my friend, actor and fitness buff Matthew McConaughey: "Just try to break a sweat every day."

Discover. A 2014 study from the University of Texas at Dallas tells us that picking up a new hobby, like painting or digital photography, or even learning a new piece of software or language can strengthen the brain.[11] Doing something new can even be seeing a 3D movie, joining a new club, or even using your nondominant hand to brush your teeth. As part of this conversation, I'll cover the benefits and pitfalls of brain training exercises, as well as how to discover your brain's full capacity through strategies that heighten attention, focus, and concentration. I will be asking, "Do you have a strong sense of purpose in life?" That, too, will be part of the equation.

Relax. Relaxing is not solely a physical thing for the body; your brain needs to chill out too. Scores of well-designed studies, some of which we'll explore in chapter 6, routinely show that poor sleep can lead to impaired memory and that chronic stress can impair your ability to learn and adapt to new situations. According to a group of researchers at MIT, something as commonplace (and stressful) as multitasking can slow your thinking.[12] Stress is particularly subversive. I'm going to help you find ways to unwind, and it won't entail mandatory meditation (though you're welcome to try it; see chapter 6). This involves both engaging in stress-reducing activities and ensuring you achieve restorative sleep on a nightly basis.

Nourish. The link between diet and brain health has long been anecdotal. But now we finally have evidence to show that consuming certain foods (e.g., cold-water fish, whole grains, extra virgin olive oil, nuts and seeds, fibrous whole fruits and vegetables) while limiting certain other foods (those high in sugar, saturated fat, and trans-fatty acids) can help avoid memory and brain decline, protect the brain against disease, and maximize its performance. Eating well is more important than ever now that we know our diet can affect our brain health (and overall health too). This conversation extends to the health of our microbial partners as well. The human gut microbiome—the trillions of bacteria that make their home inside our intestines—have a profound role in the health and functioning of our brains, and it turns out that what we eat contributes to the microbiome's physiology all the way up to our brains.

Connect. Okay, so if crossword puzzles get a B- for their ability to boost brain function, what gets an A? Connecting with others. In person and face-to-face. A 2015 study, among many others, tells us that having a diverse social network can improve our brain's plasticity and help preserve our cognitive abilities.[13] Interacting with others not only helps reduce stress and boosts our immune system; it can also decrease our risk of cognitive decline.

Get ready to reinvent the way you live. I will make this doable and practical. Your brain—no, your whole body—will love it.

PART
2

THE BRAIN TRUST

HOW NOT TO LOSE YOUR MIND

P revention is the most powerful antidote to illness, and this is especially true of degenerative maladies like those in the brain and nervous system. Shockingly, half of adults don't know the risk factors for dementia, which makes the disease even more misunderstood and scary. You can't prevent something you don't understand and cannot "see."

Age is the strongest known risk factor for dementia and Alzheimer's disease, and chronological age isn't something anyone can teach you to slow down—yet. What we know is that the incidence of Alzheimer's or vascular dementia increases exponentially after sixty-five years of age, nearly doubling every five years.[1] By the age of eighty-five and older, about a third of people have dementia.[2] But this doesn't mean that the disease takes root during these decades. Among people who are eighty-five years old, an age at which more than 30 percent have developed dementia, signs of brain decline began silently when they were between fifty-five and sixty-five years old. Similarly, the brain health of the 10 percent or so of people who are sixty-five years old and have developed dementia started to quietly degenerate when they were between thirty-five and forty-five years old. In the words of one prominent neurologist, "Alzheimer's disease may be more aptly termed a younger and middle-aged person's disease."

We don't usually think about dementia when we're entering our prime, but we should, because it provides a remarkable opportunity. Data from longitudinal observational studies accumulated over the past few decades have shown that aside from age, most other risk factors for brain disease can be controlled. That means you indeed have a powerful voice in controlling your risk for decline. As you might guess, some of the most influential and modifiable factors related to that decline are linked to lifestyle: physical inactivity, unhealthy diet, smoking, social isolation, poor sleep, lack of mentally stimulating activities, and misuse of alcohol. Half of Alzheimer's cases in the United States alone could be caused or worsened by a combination of these bad habits. High blood pressure, obesity, diabetes, and high cholesterol, especially at midlife, substantially increase the chances of developing dementia later—sometimes decades down the road. Prevention should start early, but to make it count, you need a strategy. It has to be something you can easily incorporate into your life. In Part 2, I'll give you a library of tools to adapt now that will tremendously increase the chances you can be sharp for as long as you live. These tools reflect the five pillars of preserving your brain health and function and culminate in a tailored twelve-week program.

I will also explain why these various factors have an impact on your brain so you can best understand and visualize the benefit you are getting by applying my ideas to stay sharp. Think of them as your own personal brain trust. The best part is that they are all well within your reach.

CHAPTER 4

The Miracle of Movement

*Physical fitness is not only one of the most
important keys to a healthy body, it is the basis of
dynamic and creative intellectual activity.*

JOHN F. KENNEDY

When people ask me what's the single most important thing they can do to enhance their brain's function and resiliency to disease, I answer with one word: *exercise*—as in move more and keep a regular physical fitness routine. Maybe you expected me to say diet, crossword puzzles, or higher education, but it is all about physical movement. Truth is, even if you've never maintained a consistent workout in the past, you can start today and have quick and significant effects on your brain's health (and your whole body, obviously). Fitness could very well be the most important ingredient to living as long as possible, despite all the other risk factors you bear—age and genetics included. And while it may seem hard to believe, exercise is the only behavioral activity scientifically proven to trigger biological effects that can help the brain. We cannot yet say that exercise will reverse cognitive deficits and dementia, but evidence is mounting to heed the advice that we all would do well to follow: get in motion. Remember: A body in motion tends to stay in motion. And, if you have not been exercising, starting today can significantly protect your brain later. It's never too late!

Do you know any octogenarians who can bench-press 115 pounds? I do; she lives in Baltimore and teaches a workout class at the gym. But Ernestine Shepherd didn't start exercising until she turned fifty-six, when she decided to get into shape with her sister. How about a seventy-seven-year-old ballerina (Madame Suzelle Poole) and a fifty-some professional soccer player (Kazuyoshi Miura)? In 2018, eighty-seven-year-old John Starbrook became the oldest runner to complete the London Marathon. Linda Ashmore swam the English Channel at seventy-one. These people are proof that exercise can be a lifelong activity and it's never too late to start. Scientists are finally getting around to studying "master athletes"—people who engage in sports and are thirty-five or older. They give us a wonderful glimpse into what's physically possible as we age and how exercise tangibly benefits us not only physically, but also mentally. For starters, these studies are debunking a lot of myths about the aging process. Contrary to what you might think, we don't get that much slower with age until we reach seventy. And we can gain a lot more than we previously realized from relatively low-intensity activities like walking,

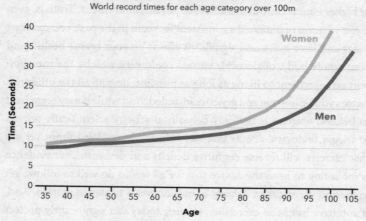

How Men and Women Get Slower as They Age

World record times for each age category over 100m

Source: World Masters Athletics records for 100-meter dash, 2019

gardening, or ballroom dancing. When I saw the chart on the previous page, the first thing that went through my mind was, "I have no excuse anymore!" It immediately put things into perspective.

THE PACE OF AGING

The influence of exercise on brain function is so spectacular that in early 2018, the American Academy of Neurology published new guidelines for doctors like me to use in deciding best choices when treating patients, particularly those with mild cognitive impairment (MCI), often a precursor to dementia.[1] The subcommittee assigned to updating the recommendations diligently reviewed eight medications that might prove helpful in slowing the progression from MCI to full-blown Alzheimer's disease. Having arrived at this point in the book, it probably won't surprise you that the panel concluded not a single drug was effective: Although there are Food and Drug Administration–approved drugs for treating the symptoms of Alzheimer's dementia, "[t]here are no FDA-approved medications for the treatment of MCI. Moreover, there are no high-quality, long-term studies identifying pharmacologic or dietary agents that either improve cognition or delay progression in patients with MCI." But what the scientists did declare is that exercise should be recommended: "Six-month studies suggest a possible benefit of twice-weekly exercise for cognition in MCI. Exercise also has general health benefits and generally limited risk." If that doesn't sound compelling enough, you should also recognize that physical *inactivity* has been calculated to be the most significant risk factor in cognitive decline and the development of dementia.[2]

Consider that while no particular medication is universally recommended, exercise is, if for no other reason than to prevent inactivity. It is an example of how the body and the brain want to heal and how movement can help make that happen. The Mayo Clinic's Dr. Ron Petersen—a founding member of the Global Council on Brain Health—was among the authors of the new guidelines. Dr. Petersen is a neurologist who has dedicated his life to studying cognition in normal aging, as well as in a variety of disorders such as Alzheimer's disease, Lewy body dementia, and frontotemporal lobar degeneration (progressive loss of nerve cells in the frontal and/or temporal lobes of the brain, causing decline in behavior, language, or movement; it's the most common form of dementia for people under age sixty). He is a world leader in the field of Alzheimer's research and directs the Mayo Clinic's Alzheimer's Disease Research Center and its Study of Aging. When I spoke with him about his thoughts on preserving brain function overall, exercise was at the top of his list. "The literature is pretty good about the role of exercise, especially aerobic exercise," he said. "Brisk walking can do it." *Walking!* It seems the basics really do apply, even when you speak to scientists at the highest levels who have dedicated their lives to studying the brain.

Dr. Petersen has seen his field revolutionized by imaging technology over his career. In his early days, doctors like him could diagnose Alzheimer's disease only on autopsy. Now special PET scans allow us to look inside the living brain and see what's going on without having to bring a scalpel anywhere close. All kinds of various imaging technologies are helping us measure changes in the brain under certain circumstances. Physical activity has the strongest evidence to date of positive brain changes. And again, it takes far less exercise than you might imagine: if brisk walking alone can accomplish the job, then there's your directive. But you have to engage in regular physical exercise at least 150 minutes a week and incorporate interval and strength training into the mix. Interval training means you alternate between varying levels of speed, intensity, and effort. Think of it as surprising the body so you don't fall into well-worn ruts that fail to challenge the body and lead to a plateau in your progress. Strength training refers to the use of weights or just your own body weight as resistance. This helps build muscle mass and tone and helps balance and coordination.

People often tell me they "don't have time" to exercise, but you must make the time. On your busy schedule, it may be the first thing to get cancelled, but it's time to change that. Remember, this isn't about vanity or looks; it's about your life and well-being. Physical exercise may offer the greatest return on investment in yourself, and it's an antidote to many things that play into your risk for decline. Here's a simple example: You already know that having high blood pressure or diabetes increases your chances of suffering from dementia later in life, but it turns out that exercise is one of the most powerful tools for gaining control of those problems too.

According to the Centers for Disease Control and Prevention, fully 80 percent of Americans don't get enough regular exercise. Only about 23 percent of men and 18 percent of women meet the recommended requirements. The people most likely to exercise are between the ages of eighteen and twenty-four (almost 31 percent of exercisers). An analysis of U.S. adults aged fifty to seventy-one found those who had exercised between two and eight hours a week from their teens through to their sixties had a 29 to 36 percent lower chance of dying from any cause over the twenty-year study period.[3]

SMARTER AND BIGGER BRAINS IN MINUTES OF MOVEMENT

I know I'm not the first to tell you about the tremendous physical healing power of exercise, but I may be the first to explain how it makes you mentally sharper. Broadly speaking, exercise improves digestion, metabolism, body tone and strength, and bone density. Most of us think about it as a weight loss tool, which it is. But it's much more than that. It can turn on your "smart genes," support emotional stability, and stave off depression and dementia. When you choose the right exercise for you, it's enjoyable and increases your self-worth and confidence. Don't take this lightly because I really mean it: You can be smarter by some measures after one hour of exercise through the effects of movement on the brain. So how does this happen?

It's not that exercise automatically injects the brain with facts about history or how to perform complex math or fly an airplane. But you will be boosting your brain in ways that help you think more quickly and clearly and be focused. This happens through both multiple direct and indirect effects, which we'll explore shortly. Try it yourself. Go out for a fast-paced walk around the block, and when you return, note how you feel and how your mind is humming. My bet is you will have more mental energy even if the walk left you out of breath. And you will probably feel more optimistic and better able to tackle the challenges of the day. Philosopher and psychologist William James stated it perfectly in the nineteenth century: "Keep the faculty of effort alive in you by a little gratuitous exercise every day."

I came to be a consistent exerciser later in life. Always more of a bookish person, I didn't equate physical exercise as anything more than a way to get fit or provide a form of recreation. I was about thirty when I started to think about exercise as a way to improve my brain. This was during a time when schools around the country were cutting back on the length of recess and physical education classes in favor of more basic curriculum courses. What partly triggered the shift was a dip in the standardized test scores in the United States, which were lagging far behind those in other countries. The prevailing sentiment became "more math and less recreation."

I began to look into the research on the impact of these sorts of policy changes on learning overall, and what I found was unmistakable: Places where students spent more time and energy engaging in individual and team sports enjoyed a positive learning impact, and places that had cut back on those had the opposite effect. It was the first time I started to think of exercise as a way of improving not only my body but my mind as well. Truth is, while anecdotal evidence going back millennia revealed benefits to exercise, it wasn't until the mid-twentieth century that large-scale studies were performed showing that physical fitness both prevented illness and protected health. Before that, it was considered mainly just a form of leisure and sport. Exercise physiology became a bona fide field of study. Now, a new study seems to emerge every week showing the neuroprotective benefits of exercise and that sedentarism (a.k.a. couch potato syndrome) appears to cause the brain to atrophy, or physically shrink, while simultaneously increasing the risk for Alzheimer's disease and other types of dementia.

To be clear, this isn't about your body habitus. Simply being inactive, regardless of your body weight, has been shown to be twice as deadly as being obese. And if you've been keeping up with the latest health news, you've likely come across headlines calling couch potatoes "smokers"—as in, "sitting is the new smoking." That's a misleading overstatement because you can't compare those two habits. The risks of chronic disease and premature death associated with smoking are substantially higher than for sitting. But the headlines nonetheless highlight an important fact: Prolonged sitting—more than eight hours a day with zero physical activity—can kill you or lead to an early death. Most of that damage is metabolic. Here is what happens. When you are immobile, your circulation slows down and your body uses less of your blood sugar, which means more sugar is circulating. Being motionless also negatively influences blood fats, high-density lipoprotein (the good cholesterol), resting blood pressure, and the satiety hormone leptin (which tells you when to stop eating). Sitting puts muscles into a sort of dormant state where their electrical activity is diminished, leading to atrophy and breakdown. Moreover, the production of lipoprotein lipase, the enzyme that breaks down fat molecules in the blood, is shut down, leading to more fat circulating as well. As your metabolic rate plummets, you stop burning as many calories.

The good news is that if you're active, even those few minutes in motion will counter the effects of being on your butt too much. The point is that while lack of exercise is a risk factor for early illness and death, simple movement itself is shown to prevent such a fate. A 2015 study out of the University of Utah School of Medicine, for instance, showed that getting up for light activity such as walking for two minutes every hour was associated with a 33 percent lower chance of dying over a three-year period.[4] Two minutes! That's a big boost in prevention for a short period of time. A mere 120 seconds each hour can offset the damaging effects that prolonged sitting has on the body.

MYTH: As you age, muscle mass is not as important as having healthy cardiorespiratory fitness.

TRUTH: People fail to appreciate how valuable muscle mass is to quality of life, recovery from illness and injury, and the ability to stay mobile and active and perform basic everyday tasks, as well as to overall metabolic health. Unlike fat, which mostly stores calories, muscle is a highly active tissue that burns calories. This helps explain why lean, more muscular people tend to burn more calories at rest than do people with higher proportions of body fat. So in addition to keeping a cardio routine that gets your heart rate up, you'll want to continue to build and maintain muscle mass. Gradual muscle loss over time goes with aging, but you can counter this decline with strength and resistance training.

MOVING THROUGH EVOLUTION

Throughout most of human history, we've been physically active every day. We had to be in order to survive. Science has even proven that over millions of years, our genome evolved in a state of constant physical challenge—that is, it took a massive amount of physical effort to find food and water. Put another way, our genome *expects* and *requires* frequent movement. I often tell my students: "We humans were not designed to sit or lie down for twenty-three hours a day and then go to the gym for an hour. Science has revealed that we humans are designed to be pretty consistently active right down to our molecular core."

Biologist and paleoanthropologist Daniel E. Lieberman of Harvard knows a lot about the power of physical activity on how the body looks and functions. His research into the evolution of *Homo sapiens* and our history of athleticism culminated in a highly referenced 2004 paper in the journal *Nature* coauthored with Dennis M. Bramble of the University of Utah.[5] They say that we've survived this long on the planet by virtue of our athletic agility. When our ancestors tracked predators and hunted down valuable prey for food, they cemented our continuing existence. We were able to find sustenance and gain energy for mating, which could then allow us to pass on our genes to the next generation of stronger, hardier humans. In his 2013 book, *The Story of the Human Body*, Dr. Lieberman makes a strong case that our epidemic levels of chronic disease today are the result of a mismatch between our evolutionary roots and modern lifestyles: "We still don't know how to counter once-adaptive primal instincts to eat donuts and take the elevator."[6] In a follow-up 2015 paper, Lieberman calls out the paradox: "Humans evolved to be adapted for regular, moderate amounts of endurance physical activity into late age," but "humans also were selected to avoid unnecessary exertion."[7] And he sums up the secrets to a good long life in the following passage from the introduction to his 2013 book, which starts with exercise: "Men and women aged forty-five to seventy-nine who are physically active, eat plenty of fruits and vegetables, do not smoke, and consume alcohol moderately have on average one-fourth the risk of death during a given year than people with unhealthy habits."[8] This should inspire you because these dictums are practical. Anyone can follow them.

It is documented that during 600 BCE—more than 2.5 millennia ago—a physician from Sushruta, an Indus Valley civilization, was the first recorded doctor to prescribe moderate daily exercise for his patients and to indicate that "it should be taken every day."[9] Sushruta recommended exercise because it made the body strong, firm, and light; supported the growth of limbs and muscles; improved digestion and complexion; prevented laziness; and *reduced senility*. English translations of the original text in Sanskrit call exercise "absolutely conducive to a better preservation of health."[10] More than two thousand years ago, the medical community recognized the link between movement of the body and the health of the brain, and once again it is starting to take center stage.

Benefits of Exercise[11]

Exercise has long been linked to positive brain health. We know that, but I want to make sure you understand how movement of your body improves your brain. A big factor is the control of blood sugar through exercise. Using sugar to fuel your muscles instead of sitting idle in your blood helps prevent dramatic glucose and insulin fluctuations that you have already learned increase the risk for dementia. Exercise also helps lower inflammation and that is critical in preventing dementia. Consider these other benefits:

- Lowered risk of death from all causes
- Increased stamina, strength, flexibility, and energy
- Increased muscle tone and bone health
- Increased blood and lymph circulation and oxygen supply to cells and tissues
- More restful, sounder sleep
- Stress reduction
- Increased self-esteem and sense of well-being
- Release of endorphins, the brain chemicals that act as natural mood lifters and pain relievers
- Decreased blood sugar levels and risk for insulin resistance and diabetes
- Ideal weight distribution and maintenance
- Increased heart health, with lower risk for cardiovascular disease and high blood pressure
- Decreased inflammation and risk for age-related disease, from cancer to dementia
- Stronger immune system

SHAPE YOUR BRAIN BY GETTING INTO SHAPE

The biology of how exercise benefits brain health goes far beyond the reasoning that it facilitates oxygenated blood flow, delivering nutrients for neural cell growth and maintenance. We've known for a long time that cerebral blood flow is a good thing. The latest science behind the magic of movement in protecting and preserving brain function, however, is worth understanding and less known among the general public. Again, there are generally two ways that exercise benefits the brain. For one, exercise effectively uses circulating blood sugar and reduces inflammation while stimulating the release of growth factors, substances that promote both the proliferation and function of cells. In the brain, these growth factors support the health of new neurons, the recruitment of blood vessels, and the survival of all neurons. The other way that exercise can benefit the brain may seem a little less objective, but it is no less important. We now know that regular movement measurably reduces stress and anxiety while improving sleep and mood—all of which can also positively affect brain structure and function. These combined effects build critically important brain resilience in the long term and help pave the way for us to be creative and insightful and to solve problems in the short term.

I have little doubt that we will soon have sufficient evidence to conclusively state that physical activity reduces the risk of dementia. We already know that people who lead a physically active life have a lower risk of cognitive decline, and research is now emerging to show that greater fitness is correlated with maintaining better processing skills in aging brains. A 2018 study, for instance, showed that the more fit an older person was, the better the chances he or she could recall words, as compared to those who were less fit.[12] I agree with what many of my colleagues like to say: Exercise can act as a "first aid kit" for damaged brain cells, speeding up recovery after injury, stroke, or a significant emotional stress. I don't know of a single pill that can do all that.

I've witnessed the merits of fitness in my own reporting and my own life. After all the years traveling the world meeting people from different backgrounds and cultures, the one pattern I've noticed is that those who are physically fit enjoy sharper minds. You have probably seen it as well. These are people whose brains don't seem to age. And for me, staying in shape is what allows me to do all that crazy traveling and at times be unreasonably productive. Exercise helps me think better and consolidate new information. Without it, I find that most of what I think are my "new" thoughts are basically a repackaging of old ideas. With my brain on exercise, I find that I am more likely to have truly novel thoughts, an incredible feeling.

Inner strength and mental toughness are often the result of controlling one ubiquitous reality of modern life: stress. When it comes to the positive effects of exercise on the brain, stress reduction is something you have likely experienced every time you work out. I've mentioned the stress-reducing effects of exercise numerous times already, but here's how it works. When your body senses stress, it releases the stress-related hormone cortisol, which is increasingly being blamed for creating long-lasting brain changes. This is why young people who are exposed to chronic stress in early childhood are prone to mental problems such as anxiety and mood disorders later in life. Daniela Kaufer is an integrative biologist at the University of California, Berkeley, and a few years ago, she and her colleagues conducted a series of experiments demonstrating that chronic stress and elevated levels of cortisol can also negatively affect memory and learning in surprising ways.[13] The scientists found that too much cortisol can generate an overproduction of myelin-producing cells, known as oligodendrocytes, and an underproduction of neurons. Think of this as having electrical wire with a lot of coating, which is the myelin, but little actual copper, the neurons, to conduct the electricity. This led to a shrinking of the hippocampus, the memory hub in our brain. Her team also found that chronic stress causes neuronal stem cells, baby precursor cells that would normally turn into neurons, to instead flip into cells that inhibit connections to the prefrontal cortex area of the brain, which is where learning and memory occur.

These are just a few examples of how stress affects the brain. Try and visualize this interplay between stress and your brain. Once you grasp this concept, you will be better equipped to control stress and its resulting flood of cortisol. Again, one of the best and simplest ways to do that is through exercise.

MYTH: Exercise can be dangerous when you get older and the body grows increasingly frail.

TRUTH: Exercise should be a lifetime activity. It will boost your brain and body in ways that can physically de-age you while preventing and even treating frailty. It is one of the most effective, drug-free methods to improve mobility and independence in older people.[14] Recent studies of recreational cyclists aged fifty-five to seventy-nine suggest they have the capacity to do everyday tasks very easily and efficiently because nearly all parts of their body are in remarkably good condition.[15] The cyclists also scored high on tests measuring mental agility, mental health, and quality of life. This doesn't mean you need to choose cycling as your go-to exercise. Choose something you like and that works with your body. If you're prone to falls or have osteoarthritis or bad knees, you'll want to pick an activity that doesn't make you more vulnerable to injury or exacerbate any condition. Swimming, for example, is an excellent way to get a great workout without any impact or risk of falling.

Earlier I referenced studies showing that people with high blood sugar—whether or not their blood sugar level technically makes them diabetic—have a faster rate of cognitive decline than those with normal blood sugar. But I haven't explained yet how this happens. High blood sugar can lead to dementia for several reasons. For starters, the condition can weaken the blood vessels and thus increase the likelihood of mini-strokes in the brain, which can then trigger various forms of dementia. Second, a high intake of simple sugars can make cells, including those in the brain, insulin resistant. That means the insulin is present but is not working as well. In turn, brain cells can't adequately absorb sugar to fuel their activity. That means no matter how much you are eating, your brain cells could still be starving.

Just as blood sugar is more easily managed when you eat right and move, so is high blood pressure, or hypertension, another important risk factor for dementia. In a 2014 study that followed thousands of Americans since the 1980s, some with and some without high blood pressure, Johns Hopkins neurologist Rebecca Gottesman revealed that having hypertension at midlife is a major risk factor for cognitive decline. And get this: The finding was independent of other risk factors such as obesity.[16]

In 2017, Dr. Gottesman published a follow-up study that showed how much certain risk factors, including high blood pressure, diabetes, and a smoking habit, have on your chance of suffering dementia later in life.[17] Smoking and diabetes were the biggest threats: Diabetes was linked to a 77 percent greater risk, and smoking in middle age was associated with 41 percent higher odds of dementia. Hypertension was associated with 39 percent greater odds of dementia. Gottesman's work has also documented that obesity can double the risk of having elevated amyloid proteins in the brain later in life.[18]

One recent study in particular that I want to point out was done in 2018. Instead of simply looking for a connection between self-reported exercise habits and brain health, this one, from the University of Texas Southwestern Medical Center, used a more precise way of measuring physical fitness.[19] The researchers decided to test the participants' maximum oxygen consumption during aerobic exercise. This is known as the VO2 max test, a method recognized by the American Heart Association as a more definitive way of evaluating cardiovascular fitness. The participants were a mix of healthy older adults and people with mild cognitive impairment. Their average age was 65.

All of the participants took a series of tests: a VO2 max aerobic test on a treadmill (similar to a cardiac stress test that lasts about ten minutes), and cognitive tests of memory and reasoning. In addition, the researchers imaged their brains with special scanning technology to see the integrity, or functionality, of their white matter—the bundles of nerve fibers through which messages pass between different areas of gray matter. We know that the health of white matter signifies how well brain areas communicate. If it begins to break down, which can happen with age, weak white matter means weaker connections throughout the brain.

The study's results highlighted an important aspect of the effects exercise has on the brain. It showed a strong association between lower levels of aerobic fitness and weaker white matter in those with mild cognitive impairment, which correlated with lower brain function. These individuals didn't do so well on the tests for memory and reasoning. In sum, the researchers connected the dots and concluded that being physically fit correlates with healthier white matter. And healthier white matter correlates with better memory and reasoning abilities. Research is underway to understand what fitness level is ideal to markedly reduce the risk of dementia and perhaps significantly slow down the progression once symptoms develop. If the simple act of moving more can slash your risk of dementia and put the brakes on any disease in motion, then there's no excuse.

JUST AS YOU WOULD BRUSH YOUR TEETH

"Exercise" includes a combination of purposeful aerobic cardio work (e.g., swimming, cycling, jogging, group exercise classes), strength training (e.g., free weights, resistance bands, gym machines, mat Pilates, lunges, squats), and routines that promote flexibility and balance (e.g., stretching, yoga). It also includes leading a physically active life throughout the day (e.g., taking the stairs instead of the elevator; avoiding prolonged sitting; going for walks during breaks; engaging in hobbies such as dancing, hiking, and gardening).

For me, exercise is a daily nonnegotiable activity like brushing my teeth. Make it the same for you. I try to break a sweat every day, aiming for about an hour of exercise in addition to as much natural movement as possible throughout the day. My go-to's are either swimming, cycling, or running, and I throw in dedicated strength training a few times a week as well. I started participating in triathlons when I turned forty because I felt that my own aging process was sapping my energy and muscle mass. I also began to worry more about the heart disease in my family that strikes men in their forties. The same old routine of occasionally playing tennis and doing some running was no longer having an impact on me. I had to build more structure into my fitness regimen and add more variety. I also made exercise a greater priority in my life. As a father of three with a demanding job and ongoing projects, I still find a way to fit exercise in every day. Human behavior dictates that you will fill whatever time you are given to complete a task, and people think of exercise as the first expendable thing when they get busy and want another hour of time for something else. I just don't do that; exercise is sacred time on my schedule.

Wherever I am in the world, I have my running shoes, swimsuit, and goggles. I take resistance bands with me as well to get in some strength training and, at the recommendation of my chair of neurosurgery, I do a hundred push-ups every day. For me, convenience is critically important. I make exercise accessible to me by having certain tools within easy reach. For example, I keep weights in my bedroom, and I've got a door frame pull-up bar at home and in my office. Incidentally, pull-ups are a great way to build your back muscles and strengthen your core. They are hard at first, but you start to feel the payoff almost immediately. People often neglect upper body strength, especially as they get older, but it's good for posture, bone density, and metabolism, and it even helps your lungs ward off pneumonia, especially if you find yourself in a hospital or bedridden.

I'll be encouraging you to get moving with a twelve-week program that you can tailor to your needs. You don't need to become a championship bodybuilder like Ernestine, buy into a gym membership, or start training for an endurance event (though I love watching people in their seventies and eighties on the racetrack). Regular exercise that gets your heart pumping and your muscles flexing is all you need to do. Ideally, and at a minimum, you'll aim for at least thirty minutes of cardio work five days a week. You'll want to get your heart rate up at least 50 percent above your resting baseline for at least twenty of those thirty minutes. Sorry, but golfing with a cart doesn't count. On the other two days, try a restorative yoga class or leisure activity like walking; don't be totally sedentary.

If you want to gain the most benefits out of exercise and lower your risk of dying prematurely, newer research points to a tripling of that 150-minute per week recommendation to a little more than an hour per day. Now that may seem like a lot, but keep in mind that this amount reflects cumulative minutes of exercise, not just gym time. The study backing this claim was published in 2015 by *JAMA Internal Medicine*. Researchers with the National Cancer Institute, Harvard University, and other institutions gathered and pooled data about people's exercise habits using six large, ongoing health surveys.[20] They ended up with information about more than half a million adults. What the researchers did to identify the relationship between minutes spent exercising and risk of mortality was to divide the people up into categories: there were those who did not exercise at all, those who worked out ten times or more the recommended amount (twenty-five hours or more per week), and everything in between those extremes. Then the researchers turned to the death records. Who died? How did the death rates sync up with the time spent exercising?

Not to their surprise, they found that the people who were at the highest risk of premature death were the sedentary ones. Next came the group that exercised a little but didn't meet the recommendations of at least 150 minutes per week of moderate exercise; yet they still reduced their risk by 20 percent. The individuals who fulfilled the guidelines had 31 percent less risk of dying during the fourteen-year period compared with those who never exercised and lived longer. The keys to the longevity kingdom, however, were bestowed on those who worked out for 450 minutes weekly. And get this: these people achieved these benefits mostly by walking. *Walking!* Compared to the people who shunned exercise entirely, they were 39 percent less likely to die prematurely. How much of those benefits pertain to brain health is yet to be determined, but I wanted to mention those figures because I find the data compelling. That's 64 minutes a day for a long, sharp life. And dare I repeat myself, those minutes can be spent in a moderately-paced stroll.

Weightlifting is important, but not enough on its own. Pumping iron does confer cognitive benefits, as some studies show among older people who just lifted weights for a year. But to gain the most benefits, and which most studies prove, you have to get aerobic through activities like jogging, swimming, bicycling, dancing, hiking, or brisk walking at least five days a week for at least twenty minutes.

I hope the evidence I've provided in this chapter will motivate you to get moving more often if you don't already maintain an exercise routine. I'll ask that you make an effort during the program to focus on this important area of your life and start a regular workout if you don't already have one. Rethink those priorities. And if you are active, then you can work on increasing the duration and intensity of your workouts or try something new. It is all a part of becoming **physically stronger and mentally sharper**.

CHAPTER 5

The Power of Purpose, Learning, and Discovery

It's not enough to have lived. We should be
determined to live for something.

DR. LEO BUSCAGLIA

The two most important days of your life: the day
you are born. And, the day you figure out why.

MARK TWAIN, OR SOMEONE ELSE

I will probably never retire. I wouldn't know what to do with myself. I also know the consequences for people who retire early: an increased risk of developing dementia. They also are more likely to suffer from other conditions that further increase that risk, including depression. One study showed that for each additional year of work, the risk of getting dementia is *reduced* by 3.2 percent.[1] The study, which included nearly half a million people in France, showed that someone who retired at age sixty-five had about a 15 percent lower risk of developing dementia compared to someone retiring at sixty, even after other factors were taken into account. (France has produced some of the best Alzheimer's research in the world, partly because its former president, Nicolas Sarkozy, made it a priority. Part of their strides in this realm has been because the country keeps detailed health records on self-employed people who pay into a Medicare-like health system, so there's more data to mine.)

The conclusion makes sense. Staying engaged in a job, especially one that's satisfying, tends to keep people physically active, socially connected, and mentally challenged—all things known to protect cognition. Years ago, while I was on a search to find the secrets to longevity for a project, I spent a lot of time in Okinawa, Japan. There isn't even a word for retirement in Okinawa. People do different things as they get older, not necessarily fewer things. They are also honored and included even more as they age, as a sign of respect but also an acknowledgment of their experience. My visits to Okinawa over the years left a profound impression on me, and I am pretty sure their approach is how I want to grow old.

The lesson: delay retirement as long as possible. And when you do retire, don't quit on life. Find activities that are joyful and stimulating. Stay engaged. There's power to maintaining a sense of purpose by continuing to learn, discover, and complete complex tasks. A sense of purpose means that you see your life as having meaning, a sense of direction, and goals to live for. It's active aging.

KEEPING THE BRAIN PLASTIC

As you might have guessed, active aging involves more than moving your body. You also need to move your brain, exercising it in ways that keep it healthy. Recruiting your muscles in exercise improves overall health; using your brain in challenging ways similarly improves overall brain health. But there is a right way and a wrong way to employ the brain. Pick the right way, and it will help you tap the "plastic" power of the brain—its capacity to rewire itself and strengthen its networks.

One of the most startling pieces of research has been comparing different people's brains when they are autopsied. I know it's not for everyone, but participating in an autopsy of someone's brain is one of the most enlightening experiences I have had. You get to see deep into this mysterious organ in ways you couldn't when the organ was alive. And one of the big revelations is that while some brains may share nearly identical pathology, their owners sometimes exhibited very different behaviors when they were alive. Two brains that look severely diseased at autopsy, perhaps riddled with the plaques and tangles of Alzheimer's disease or signs of cerebrovascular disease, will not necessarily reflect how their owners came across in life. One person may never have shown any symptoms of cognitive impairment or decline, whereas the other faded away for years and could not recognize any family members' faces toward the end of life. The question I always asked was how the person with a seemingly sick brain averted the cognitive decline. The answer I often heard was "cognitive reserve" or what scientists call *brain resiliency*. Building that reserve or resiliency has everything to do with staying as engaged in life as possible through socialization and participating in stimulating activities. I'll cover the importance of connecting with others in chapter 8. For now, let's focus chiefly on the concept of cognitive reserve. Think of it as a big backup system in the brain that results from enriched life experiences such as education and occupation. You will learn that cognitive reserve may even help counteract the effects of other risk factors such as a poor diet.

THE BRAIN AND COGNITIVE RESERVE

The whole idea of a cognitive reserve, or brain resiliency, remains a bit controversial because we're not sure exactly how it works and it can be difficult to define. From a practical standpoint, cognitive reserve is your brain's ability to improvise and navigate around impediments it may encounter that could prevent it from getting a job done. To draw on another car analogy, your car has a breaking and acceleration system for navigating the road and dealing with things it may encounter, like obstacles and unforeseen turns. You can quickly swerve to avoid an accident and stay on course. Similarly, your brain can change how it operates to find alternative routes, thereby helping it to cope with challenges that could otherwise be harmful to its health and function. If you think of your brain's networks like a series of roads, then you can see how the more networks you have, the more options are available to shift direction and arrive at the same destination if one road becomes impassable. It's a simple way of looking at this, but those networks or roads are the cognitive reserve, and they develop over time through education, learning, and curiosity. The more you discover in your lifetime, the more networks you create to help your brain better manage any potential failures or declines it faces.

The concept of cognitive reserve is relatively new. It originated in the late 1980s when a group of scientists in the Department of Neurosciences at the University of California San Diego described older people in a skilled nursing facility with no apparent symptoms of dementia who were nonetheless found at autopsy to have brains whose physical appearances were consistent with advanced Alzheimer's disease. Their paper, published in the *Annals of Neurology*, was the first to use the term *reserve*, suggesting that these individuals had enough brain cache to offset the damage and continue to function as usual.[2] The researchers also noted that the people who had escaped symptoms of dementia had higher brain weights and a greater number of neurons.

Since that revolutionary finding, research has consistently shown that people with greater cognitive reserve are more likely to stave off the degenerative brain changes associated with dementia or other brain diseases, such as Parkinson's disease, multiple sclerosis, or stroke.[3] A more robust cognitive reserve, the researchers say, can also help you function better for longer if you're exposed to unexpected life events that can impact the brain, including chronic stress, surgery, or toxins in the environment. These types of circumstances demand extra effort from your brain, just like a car needs to engage another gear to deal with the demands of a steep hill. Two forms of cognitive reserve are often discussed: neural reserve and neural compensation. In neural reserve, preexisting brain networks that are more efficient or have greater capacity may be less susceptible to disruption. In neural compensation, alternate networks may offset or balance out any disruption of preexisting networks.

So an important goal is to build and sustain your cognitive reserve, and this can be done by maintaining demands on your brain that keep it thinking, strategizing, learning, and solving problems. This is not something you do overnight. Cognitive reserve is a reflection of how much you have challenged your brain over the years through your education, work, and other activities. It is the reasoning behind why epidemiologic evidence suggests that people with higher IQ, education, occupational achievements, or engagement in leisure activities—participating in hobbies or sports that are not related to a job—have a reduced risk of developing Alzheimer's disease. These pursuits force the brain to continually acquire knowledge and work with that knowledge in ways that ultimately build new networks and strengthen existing ones. Not surprisingly, animal studies show that cognitive stimulation increases the density of neurons, synapses, and dendrites. Put more simply, cognitive stimulation builds a brain more resistant to disease.

To say that having a higher IQ and advanced degree will help protect you from dementia is not to suggest that being "smarter" or more degreed will stave off disease. That's not the point. In fact, the long-held theory that a college education will fend off dementia later in life has been debunked as a result of a 2019 study published in the journal *Neurology*.[4] The nearly 3,000 participants in the study were around seventy-eight years old when they enrolled in it. They had on average 16.3 years of education and were followed for eight years. Nearly 700 of the participants developed dementia during the study; 405 developed dementia and also died, while 752 died and had a brain autopsy.

Dividing the participants into three education levels, researchers did find that those with more education scored higher on tests of thinking and memory skills at the start of the study, even though it was decades after receiving their college degrees. However, the researchers did not find an association between higher education and slower cognitive decline, nor did higher education seem to delay the onset of dementia.

As study author Robert S. Wilson, director of cognitive neurosciences at Rush University Medical Center in Chicago, described the findings, "These results did not show a relationship between a higher level of education and a slower rate of decline of thinking and memory skills or a later onset of the accelerated decline that happens as dementia starts."[5] A good explanation for why higher education may not have an impact on cognitive reserve as much as once was believed is that the schooling occurs decades before the slow creep of dementia begins. In other words, you can't bank on your college or graduate degree to save you if you haven't been keeping up with your "continued education" in the form of reading, learning, and socializing. Again, when it comes to memory and aging, a use-it-or-lose-it concept applies. In that sense, this research is encouraging. According to Sarah Lenz Lock, executive director of AARP's Global Council on Brain Health, "This study suggests that anyone can work on improving cognitive reserve at any age, no matter their previous education levels."[6] Remember that new brain cell growth can happen even late into adulthood, and the brain remains plastic throughout life.

Whenever you hear about studies like this, you have to consider them in a broader context. While lifelong education appears very protective against dementia, we also know that it—whether formal or not—is a luxury, typically available to those who have better economic status, career standing, and social interactions. Teasing out which protective factors have the greatest impact and how they interact with one another is challenging. For now, the guidance is to focus on lifelong education as much as possible. It's how you continue to build and maintain that resilience in the brain I mentioned. There's something to be said for the stereotype of "letting your brain go to mush" when you don't stimulate it by learning new things and challenging its thinking and calculating abilities. For many people, the easy act of taking a book out of the library and reading it qualifies as a form of education. You don't need to go after a PhD.

THE DEFINITION OF "COGNITIVELY STIMULATING" ACTIVITIES

Unfortunately, most people get it wrong when it comes to defining cognitively stimulating activities. A large majority of Americans age fifty and older (92 percent) think that challenging the mind with games and puzzles is important to maintaining or improving brain health; a majority (66 percent) also believe that playing online games designed for brain health is the best way for you to maintain your brain health.[7] But the evidence doesn't support this. Commercial claims to the benefits of playing "brain games" are everywhere, but they are often exaggerated and can distract one from engaging in the kinds of activities that are truly cognitively stimulating. Any product that says it can reduce or reverse cognitive decline should be met with caution. In recent years, the Federal Trade Commission has aggressively cracked down on deceptive advertising from companies claiming their brain-training programs can protect against dementia and age-related cognitive decline.

Brain-training videos and games such as puzzles and crosswords can improve working memory—the ability to remember and retrieve information, especially when distracted. But research has found that although they can help your brain get better at performing those specific activities, their benefits do not extend to other brain functions like reasoning and problem solving, both of which are key to building cognitive reserve. There is also a reason why taking a traditional class can beat an online brain-training program. Classes offer a level of complexity that has long-term benefits; they not only employ cognitive skills, such as visual comprehension, short- and long-term memory, attention to detail, and even math and skills, but they often entail a social element with fellow classmates. Students in a classroom setting are interacting and communicating with others on a regular basis through lively conversation.

That doesn't mean taking a class has to be in a traditional academic setting or even entail acquiring another degree. It can simply be about learning new skills, such as speaking a foreign language, learning how to cook or paint, or taking up a new musical instrument. You can study how to computer code, take salsa dancing, or write a novel—whatever gets you out there and acquiring new knowledge and aptitudes. Just be sure that you do something you enjoy. Don't sign up for a Civil War history class if that does not appeal to you. Use the opportunity to learn more of what are you interested in now or wish you had explored before.

Research has long shown that new knowledge, whatever it is, pays off. For example, a study in the June 2014 issue of *Annals of Neurology* found that speaking two or more languages, even if you learned the second language years or decades after the first, may slow age-related cognitive decline.[8] Such findings have been confirmed by others, including cognitive neuroscientist Ellen Bialystok, a distinguished research professor of psychology at York University in Toronto, Canada. Her own research has found that bilingualism can protect older adults' brains, even as Alzheimer's is beginning to affect cognitive function.[9] It's likely that the complexity of the second language acts as part of that cognitive reserve, shielding against symptoms of decline. And therein lies a key secret: The complexity of the new skill is critical; you can't just come to class and be passive. You need to use your mind in a manner that gets you out of your comfort zone and demands more long-term memory.

Although video-based brain games have come under fire for being overly hyped, some are being further investigated and developed because certain types are showing promise. The type that's gotten the most attention lately is speed training. If you ever played Punch Buggy in your youth, then you've already experienced a mild form of speed training. Punch Buggy, or "Bug," was a popular game when I was growing up, often played by kids in cars (this was long before digital screens became driving companions). The goal of the game was straightforward: At the sight of a Volkswagen Beetle, you'd punch your fellow passenger (usually a sibling) and accumulate points. The person who spotted the most Bugs wins. Although it's very basic and juvenile in nature, the game often required visually scanning the other side of the highway and quickly filtering through all the cars to spot a Bug and be the first to call it out. This type of mental exercise, which demands that you focus intently and rapidly process visual information, appears to be surprisingly effective at putting off dementia. Speed training games have since gotten much more sophisticated, digital, and worthy of serious research.

In 2016, a secondary analysis of an original ten-year study funded by the National Institutes of Health showed that speed training was more effective than memory and reasoning exercises in terms of its potential effects on reducing risk of developing dementia (the results were first presented at the Alzheimer's Association International Conference in Toronto that year and published formally in 2017).[10] A total of eleven to fourteen hours of speed training was demonstrated to potentially cut that risk by 29 percent. The primary study, called ACTIVE (Advanced Cognitive Training in Vital Elderly), was led by researchers at the Institute on Aging and six research universities across the country. It was originally designed to measure people's cognitive function and their ability to maintain basic activities of daily living. It enrolled 2,802 healthy older adults (average age at the start was seventy-four) and randomly assigned them to a control group or one of three interventions: (1) a group that received instructions on reasoning strategies, (2) a group that got instructions on memory strategies, or (3) a group that was given speed processing training with the help of computer video games specifically designed for this purpose. These games call for highly focused visual attention, even in the face of a distraction, to perform a certain task. For example, in the Double Decision game, a player had to distinguish between two blue cars—one a hardtop and one a convertible—in a progressively complex and visually distracting setting. The player might also be asked to find other visuals such as a Route 66 sign. As the player answered correctly, the game became even more complicated and mentally strenuous with more distracters so the targets become more difficult to identify. At the same time, the speed of the presentation ticked up a notch.

The speed training group was given ten initial sessions (sixty to seventy-five minutes per session) over the first six weeks of the study. All groups were assessed for functional decline using an array of cognitive and functional tests at the beginning of the study and again at intervals over the ten years. Some people also received "booster" training sessions after the first year and three years into the study. In the end, not only did the speed training group gain the most benefits, but the benefits were "dose related": those who completed more training sessions benefited more.

The secondary analysis did have its limitations, and the researchers acknowledged that the results they found related to lower dementia risk could be due to reverse causation—meaning there may not be a definitive, direct cause and effect between speed training and lower risk for dementia. Nevertheless, I think these types of explorations hold much promise. Just ask Kathy Lasky, a woman in her seventies who tried to retire several years ago from her job as a pharmaceutical technician but decided after a few months that not working wasn't for her. I interviewed her in San Diego for my show *Vital Signs* in 2017, and her story has stuck with me. "Daytime TV gets very old," she told me. Kathy was in great shape physically, but she quickly noticed that she was starting to feel mentally foggy while in retirement. Fearful that she'd sink into depression or even develop dementia, Kathy went back to work and enrolled in the ACTIVE study where she was assigned the speed training exercises. It was the twin forces of work and mental play here that likely made a difference. Today she feels as vital as ever and continues to work and participate in speed training exercises using video games. She calls brain fitness "hot sauce for the mind." And her experience in the gaming world may soon reflect a paradigm shift in brain medicine. Researchers are coming to realize that there's a largely unlocked potential for video games to train our brains to be faster, stronger, better—if those games are developed correctly.

Dr. Adam Gazzaley is a neuroscientist and inventor who knows what it means to stimulate the brain to improve its function and physiology. He is the founder and executive director of Neuroscape, a center at the University of California, San Francisco, that translates brain science into practical solutions, technologies, and treatments for people to optimize brain function. Dr. Gazzaley is a professor of neurology, physiology, and psychiatry at UCSF. He is also cofounder and chief science advisor of Akili Interactive Labs, a company developing therapeutic video games to support treatment of brain disorders such as attention deficit hyperactivity disorder (ADHD), autism, depression, multiple sclerosis, Parkinson's, and Alzheimer's disease. Moreover, he is chief scientist at a venture capital firm investing in experiential technology to improve human performance. His dream? To one day see doctors write prescriptions for video games approved by the Food and Drug Administration rather than pills to turn an aging brain into a younger one.

MYTH: Playing video games will make your mind turn to mush.

FACT: People who play video games can see more than the rest of us, on average. They make better and faster use of visual input, as Duke University researchers have demonstrated.[11] The video gaming industry is poised to explode as we learn more about how to design games to improve brain health and function.

Dr. Gazzaley is considered a maverick in brain optimization and a pioneer in digital medicine. He's defining the difference between what truly works on the brain to improve its performance and potentially stave off decline and what's hype. He appreciates the power of programs that move the mind. Using the latest technologies to visualize the brain's functionality in real time, such as functional and three-dimensional MRI and electroencephalograms, Gazzaley is able to watch and document changes in the brain as it's stimulated in various ways—notably by video-based brain games that demand concentration, hand-eye coordination, and avoidance of distractions. He hooks willing participants up to one of these advanced brain-imaging technologies, hands over a gaming control, and lets them play. Then he captures their brains' activity—identifying which areas light up and gain increased electrical activity. These kinds of experiments were unheard of just a few years ago. We've come a long way since Pong came on the scene in 1972 and Tetris reigned king in the 1980s.

When I met up with Gazzaley at his lab at the Center for Integrative Neuroscience at UCSF, I had the pleasure of witnessing his research in live action with individuals plugged into his revolutionary brain model, Glass Brain, a computerized simulation of a person's brain that shows exactly what's going on in the instant someone is playing a game and being mentally, and sometimes physically, challenged. It paints a wild and vivid picture of all the signaling that's taking place in the moment. He gets to see (as did I) where the brain is firing and how strongly, and correlate that with knowledge about what those areas of the brain mean for us neurologically. "We focus on attention processes—how we direct our limited resources where and when we want them," he tells me. "When these abilities decline, we see all sorts of conditions arise, from ADHD and depression to autism and even Alzheimer's disease." Gazzaley has spent the past several years building his Glass Brain by figuring out how to challenge the brain in just the right way. For a neuroscientist like me, it was thrilling to go behind the scenes in his highly proprietary lab. I felt that I was watching the origins of groundbreaking brain medicine. I now appreciate video games from a whole different perspective, as they may very well soon become medical devices.

"Experience drives plasticity in brain," Gazzaley reminds me. "Based on the facts of neuroplasticity, we can create experiences targeted and strong enough to make meaningful change in the brain to improve and protect brain function." His work has not gone unnoticed. It was first highlighted in the journal *Nature* in 2013, in which he reported on one of his studies showing that if a game is designed to address a precise cognitive deficit—in this case, multitasking in older people—it can be effective.[12] What was incredible was finding that after participants played NeuroRacer three times a week for a month, they improved their ability to multitask beyond the level of even twenty-year-olds who played for a single visit. And the improvements were shown lasting six months later without practice. His team conducted a series of cognitive tests on the participants before and after training. Certain cognitive abilities that were not specifically targeted by the game improved and remained that way, such as working memory and sustained attention. These skills are important for performing daily tasks, such as handling mail and bills, and planning and cooking meals.

Gazzaley would agree that we should not overpromote the power of gaming to improve cognition. Video games will never be a guaranteed panacea, and there will continue to be unscrupulous players in the field selling video games with false claims. When I asked Gazzaley about "the one thing" everyone can do to preserve brain function and prevent neurodegenerative decline, his advice will sound familiar to you: "Lead a rich, active, dynamic, complex life." I can't argue with that! Gazzaley has multiple games in development that are undergoing the rigors of clinical trials; he's hopeful that FDA-approved games will one day be on the market and just as important as any drug.

A STRONG SENSE OF PURPOSE

My mother, Damyanti, is one of my heroes. She has always lived with a sense of purpose and has worked hard to instill that into me and my younger brother. My mother's drive was born out of misery. At the age of five, she was forced to flee an area of the world that is now Pakistan. It was during the time of the bloody Indian subcontinent partition. Along with her family, my mother joined one of the largest human migrations in history. After arriving in India, she lived as a refugee for the next several years, struggling to survive. People in those refugee camps didn't have the luxury of hopes, dreams, and aspirations. Yet her mother (my grandmother), Gopibai Hingorani, a woman who had completed only the fourth grade, told her she was going to make sure her daughter received something that no one could ever take away from her: an education.

It still gives me shivers to imagine a young girl trapped in a camp being told she would one day become someone who mattered. By keeping her promise, my grandmother initially gave my mother her sense of purpose. My mom completed engineering college in India and made history as the first female engineer there. It was just the beginning of her life in a male-dominated space. After reading a biography of Henry Ford, she dreamed of working for the company that he'd built. Again, my grandparents came through. They took their savings of a lifetime to send my mom to the United States in 1965. At age twenty-four, she became the first woman hired as an engineer at Ford Motor Company.

My parents are now retired in Florida, but they stay active, playing a lot of bridge, singing karaoke, and traveling. My mother spends a lot of time with her five granddaughters, teaching them the value of a life lived with purpose. Because of my parents, I began studying the objective value of purpose from a medical perspective. Over the past two decades, dozens of studies have shown that older people with a sense of purpose in life are less likely to develop a slew of ailments—from mild cognitive impairment and Alzheimer's disease, to disabilities, heart attacks, and strokes. And they are more likely to live longer than people without this strong undercurrent. In fact, feeling you have a purpose in life right now might reduce your risk of future dementia by up to 20 percent. Some of the research is eye-opening. In 2017, *JAMA Psychiatry* published a study out of Harvard revealing that older adults with a higher sense of purpose tend to retain strong hand grips and walking speeds.[13] That may sound like an odd thing to measure, but these characteristics have long been indicators of how quickly people are aging. You'd be surprised by the correlation between how fast you can walk and how fast you are aging. Another great predictor of health also happens to be whether you can get up from the floor without using your hands to prop yourself up.

The explanation for the power of purpose makes sense. With purpose comes the motivation to remain physically active and take better care of oneself. And these in turn help people to manage stress and be less prone to dangerous inflammation. We also know from autopsies performed on adults in their eighties that those who felt their lives had meaning suffered far fewer microscopic infarcts, which are small areas of dead tissue that result from a blockage of blood flow.[14] These infarcts boost the risk of a stroke and developing dementia.

Having a sense of purpose will also help keep your brain plastic and preserve that cognitive reserve. With purpose comes a love for life and all the experiences it offers. Purpose also puts a damper on depression, which can be common in one's later years and is a huge risk factor in itself for memory decline, stroke, and dementia. I should add that *ikigai* is a word heard a lot in Japan, especially in Okinawa, where certain populations have incredibly low rates of dementia. Roughly translated, it means your reason for being. I think of it as the thing that makes me want to jump out of bed in the morning. We would all do well to define our ikigai, because it is a daily reminder of our purpose here on earth. And we can't forget that with a sense of purpose comes optimism. In 2018, a Global Council on Brain Health report stated that optimism is among the important elements of mental well-being, alongside things like self-acceptance, vitality, and positive relationships.[15]

GETTING IN THE FLOW

There's no shortage of ways to stay engaged and maintain a sense of purpose. As this chapter has already pointed out, you don't have to keep a regular job. You can enroll in a class to learn something new, volunteer, teach, renew your library card, work on your hobbies, be a good friend to your neighbors, turn your garden into a sanctuary—whatever you find joyful, satisfying, and meaningful. It's also important that you find things to do that put you in the "flow." For more than four decades, social theorist Mihaly Csikszentmihalyi (pronounced MEE-high, CHEECH-sent-mee-high) has studied the concept he named *flow*, which has become a pillar of positive psychology research.[16]

We've all experienced being "in the moment," "in the groove," or "on fire." *Flow* is the word used to describe this phenomenon. It means you're in a mental state that has you totally immersed in an activity without distraction or any sense of agitation whatsoever. You're deeply focused, enjoying a feeling of intense energy as you're absorbed in the activity. You're not necessarily stressed; rather, you can feel blissfully relaxed while at the same time being challenged or "under the gun." The concept of flow has been recognized across many fields, including occupational therapy, the arts, and the sports world. Mihaly Csikszentmihalyi may have given us the popular term in modern times, but the concept of flow has existed for thousands of years under other guises, notably in some Eastern religions.

You can't truly be in the flow without a clear sense of purpose. Think about the last time you were in the flow. What were you doing? How long has it been since that time? Who were you with? I encourage you to write down those experiences. They may inspire you to find new roads to flow today.

CHAPTER 6

The Need for Sleep and Relaxation

Even a soul submerged in sleep is hard at work
and helps make something of the world.

HERACLITUS

How well did you sleep last night? Do you remember dreaming? Did you sleep solidly without waking? Do you rely on an alarm to wake you up? If you can't call yourself a good sleeper, you're not alone. Fully two-thirds of us who live in the modern, developed world are chronically sleep deprived. That's tens of millions of us. As I mentioned in part 1, I sorely underestimated the value of sleep for far too long and wish I could gain back all those hours—years, probably—that I lost. Now I rank sleep close to the top of my list in terms of priorities.

The subject of sleep has produced lots of bad information. People who tell you they can get by on four hours of sleep do not know what they are talking about.* And if they do get only that much sleep, they are

*A very small percentage of people have what's called a short sleep gene, a rare mutation in a gene that reduces their need for sleep. These individuals naturally sleep for just four to six hours and function normally. But we don't have long-term data on this phenomenon, and the vast majority of people are not genetically equipped to be short sleepers, even if they "train" themselves to wake up early.

living with a much higher risk for all sorts of health challenges.[1] Chronic inadequate sleep puts people at higher risk for dementia, depression and mood disorders, learning and memory problems, heart disease, high blood pressure, weight gain and obesity, diabetes, fall-related injuries, and cancer. It can even trigger biases in behavior, causing you to focus on negative information when making decisions. Lack of sleep is not a badge of honor or sign of integrity. If you think rising at 4:00 a.m. after going to bed at midnight will make you more successful, think again. There is no data that shows successful people get less sleep, despite the trend among celebrities and entrepreneurs to extol the virtues of their super-early mornings. You cannot game your body clock. Once you learn how important sleep is in your life, my hope is that you begin to prioritize it. We all need seven to eight hours nightly, and yet on average, Americans sleep less than seven hours a night—about two fewer hours than they did a century ago. Dr. Matthew Walker, a professor of neuroscience and psychology at the University of California, Berkeley, is among today's pioneering researchers in the power of sleep.[2] He used to say that sleep is the third pillar of good health, alongside diet and exercise. But given his latest finding about how sleep supports the brain and nervous system, he now teaches that sleep is the single most effective thing we can do to reset our brains and bodies, as well as increase a healthy life span. How could something we spend about twenty-five years of our lives doing be useless?

Contrary to popular belief, sleep is not a state of neural idleness. It is a critical phase during which the body replenishes itself in a variety of ways that ultimately affect every system, from the brain to the heart, the immune system, and all the inner workings of our metabolism. It is normal for sleep to change with age, but poor-quality sleep with age is not normal. While sleep disorders such as sleep apnea and early waking become more common with age, they often can be treated with simple lifestyle changes to improve sleep.

Sleep apnea, which affects millions of people, is caused by a collapse of the airway during sleep; muscles in the back of the throat fail to keep the airway open. This results in frequent cessation of breathing, which causes sleep to be fragmented. Dreamless sleep and loud snoring are telltale signs of the condition. Sleep apnea can be treated, usually with the help of a continuous positive airway pressure (CPAP) device worn during sleep. Because extra weight can also exacerbate sleep apnea, people who are overweight and lose weight often find relief and may no longer need a CPAP device.

MYTH: The body shuts down during sleep. Losing a little sleep is not a big deal, and even when you do, you can catch up over the weekend.

TRUTH: Sleep is anything but a waste of time. It's when the body heals tissues, strengthens memory, and even grows. Losing sleep will have both short- and long-term consequences on your health, and you cannot necessarily catch up on sleep later on by sleeping in over the weekend or taking a long, sleepy vacation.

SLEEP MEDICINE

The subject of sleep and the reason for its existence remained a mystery until the previous few decades. Sleep medicine was unheard of a few generations ago, but today it's a highly respected field of study that continues to clue us into the power of sleep in the support of health and mental wellness. If sleep were not important, so many creatures wouldn't do it; even the simplest creatures, including flies and worms, need to sleep. But we mammals appear to be particularly dependent on it. Rats that are forced to stay awake die in about a month, sometimes within days.

The quality and amount of sleep you get has an astonishing impact on you. Your body does not momentarily press the pause button during sleep. It's more like a reset button because sleep is a necessary phase of regeneration. Billions of molecular tasks go on during sleep at the cellular level to ensure that you can live another day. Sufficient sleep keeps you sharp, creative, attentive, and able to process information quickly. Studies have convincingly proven that sleep habits ultimately rule everything about you—how big your appetite is, how fast your metabolism runs, how strong your immune system is, how insightful you can be, how well you cope with stress, how adept you are at learning, and how well you can consolidate experiences in your brain and remember things. Banking six or fewer hours for a single night reduces daytime alertness by about a third and can even impair your ability to operate a car or other machinery.

Several years ago I met with Dr. William Dement at Stanford University's Sleep Research Center, which is part of the medical school there. He is fondly known as the father of sleep science. He began studying sleep in the 1950s when few people realized how much there was to discover. He quickly learned that sleep is complicated, with a lot of unknowns. In summer 1970, he opened the first sleep disorders clinic and sleep laboratory in the world to study sleep and treat the number one problem for his patients: obstructive sleep apnea (OSA). This occurs when tissues in the back of the throat collapse, blocking the airway. It's caused by excess weight, large tonsils, or just the structure of one's throat. For ten seconds to a minute or more, a person with sleep apnea stops breathing, which lowers blood-oxygen levels and strains the heart. These micro-awakenings can go on hundreds of times a night, fragmenting sleep and preventing a person from experiencing all cycles of sleep that include the most restorative one: deep sleep. OSA is incredibly common today, affecting approximately 20 percent of U.S. adults. But of these people, as many as nine in ten are undiagnosed, according to the American Academy of Sleep Medicine.[3] It's more prevalent in people over the age of fifty and in men (affecting 24 percent of men compared to 9 percent of women). The condition can increase the risk of developing heart disease, diabetes, stroke, and cancer. It also increases the risk of car accidents and lowers quality of life in general largely due to daytime exhaustion and lack of energy. Treatments are available, but the key, of course, is to get diagnosed.

Dr. Dement has since studied all aspects of sleep, from the importance of adequate slumber to the dangers of sleep deprivation. His accomplishments have paved the way for modern sleep research that can dig into what's really going on inside the brain when we close our eyes and let go. For example, one aspect to sleep that is underappreciated and uniquely influential to our sense of well-being is its control of our hormonal cycles. Each one of us, men and women, has a circadian rhythm that includes our sleep-wake cycle, the rise and fall of hormones, and the fluctuations of body temperature that all correlates with the solar day. It repeats roughly every twenty-four hours, but if your rhythm is not synchronized properly with the solar day, you don't feel 100 percent. If you've traveled across time zones and experienced jet lag, then you know—often painfully—what it means to have a disrupted circadian rhythm.

Your circadian rhythm revolves around your sleep habits. A healthy rhythm directs normal hormonal secretion patterns, from those associated with hunger cues to those that relate to stress and cellular recovery. Our chief appetite hormones, leptin and ghrelin, for example, orchestrate the stop and go of our eating patterns. Ghrelin tells us we need to eat, and leptin says we've had enough. Ever wonder why you suddenly feel hungry right before you go to bed? It makes no sense biologically because you are about to sleep. That is likely a circadian rhythm that is out of sync. The science that has made these digestive hormones so popular lately is breathtaking: we have data now to demonstrate that inadequate sleep creates an imbalance of both hormones, which adversely affects hunger and appetite. In one well-cited study, people who slept just four hours a night for two consecutive nights experienced a 24 percent increase in hunger and gravitated toward high-calorie treats, salty snacks, and starchy foods.[4] This is probably due to the body's search for a quick energy fix in the form of carbs, which are all too easy to find in processed, refined foods. And we all know what increased intake of refined carbs can lead to: weight gain. That excess weight will then haunt your metabolism and raise the risk of brain decline.

Entire books have been written on the value of sleep, but here I'm going to outline specifically the importance of sleep on brain health and function.

MYTH: The older you get, the less sleep you need.

FACT: Although our sleep patterns change as we age—we tend to have a harder time falling asleep and more trouble staying asleep than when we were younger—our sleep needs remain constant throughout adulthood.

A WELL-RESTED BRAIN IS A HEALTHY BRAIN

Early explorations on sleep initially looked into its impact on memory. In the early twentieth century, Cornell University psychologists John G. Jenkins and Karl M. Dallenbach were among the first scientists to experiment and write about the role of sleep in improving memory. Back then, we didn't really know if sleep had anything to do with memory, but these prescient researchers set out to test and quantify sleep's relationship with how we remember. Recruiting unsuspecting students for their experiment, they gave the students lists of nonsense syllables to memorize either in the morning or just before going to bed. I say "unsuspecting" because the students had no idea what goals and questions were driving the experiment. They were tested for their memory of the lists one, two, four, or eight hours later. When the students learned the lists at night, the time between learning and then having to recall the syllables was spent in sleep; in the other case, the people were awake during the intervening period. Who do you think fared better in the recall of syllables? Answer: the group that slept between the tests. It may be better described as having a slower rate of forgetting. This study has been repeated through the years in many different ways. Jenkins and Dallenbach's seminal 1924 paper in the *American Journal of Psychology* set the stage for future research that continues to this day. (Interesting tidbit: The term *oblivescence* is used to describe the process of forgetting, in this case during sleep; "oblivisi" means "forget.")[5]

Scientists have proposed several pathways for how sleep deprivation seems to induce a nearly universal "brain fog" that makes it hard for us to concentrate or remember important facts. One of the most recent theories about memory and sleep suggests that sleep helps us triage important memories to ensure we encode the most significant events in our brains. Sleep is essential for consolidating our memories and filing them away for later recall. Research is showing that brief bursts of brain activity during deep sleep, called sleep spindles, effectively move recent memories, including what we learned that day, from the short-term space of the hippocampus to the "hard drive" of our neocortex.[6] Sleep, in other words, cleans up the hippocampus so it can take in new information that it then processes. Without sleep, this memory organization cannot happen. More than just affecting memory, a sleep deficit prevents you from processing information in general. So not only do you lack the ability to remember, you cannot even *interpret* information—to bring it in and think about it.

Could sleep loss be setting us up for irreversible memory issues? That's a good question, one that science is finally addressing. An alarming 2013 study found that older adults who experience fragmented sleep are more prone to develop Alzheimer's disease.[7] Their rate of cognitive decline was also higher than that of people who routinely got a good night's sleep. Although we've known that chronic bad sleep is commonly associated with neurodegenerative diseases like dementia, recent data show us that this problem can occur years before a person is even diagnosed. In other words, sleep problems could be an early warning sign. And getting enough sleep now can improve your chances of fending off dementia in the future.

Sleep deprivation causes a number of other problems, all of them related. A 2017 paper published by the American Heart Association showed that in people who had suffered a sudden reduction or blockage of blood flow to the heart (usually due to a blood clot in the heart's arteries or a plaque rupture), fewer than six hours of sleep was associated with a 29 percent higher risk of having another major coronary event.[8] Another 2017 study, this one of eighteen thousand adults, revealed that logging fewer than six hours of sleep a night was associated with a 44 percent increased risk for those with prediabetes going on to develop full-blown diabetes; getting less than five hours a night increased the risk by 68 percent.[9]

This is key information because of the well-documented relationship between diabetes and brain health. You'll recall from part 1 that people with type 2 diabetes can have a much higher rate of cognitive decline than those who do not have the condition (and who can maintain normal blood sugar levels). As I also mentioned, this has led some scientists to refer to Alzheimer's as a type of diabetes. When the body's insulin system is broken and neurons in particular cannot use insulin properly to fuel their metabolism, the stage is set for decline.

Finally, chronic inflammation also plays a role. We still have a lot to learn about the specifics of the relationship between sleep and inflammation, but a strong body of evidence already shows that lack of sleep raises levels of inflammation. This has been shown with acute sleep deprivation, as in not getting any sleep for a full twenty-four-hour period, and partial sleep deprivation—recurring insufficient sleep that many of us experience in our nightly lives. A single night of inadequate sleep is enough to activate inflammatory processes in the body, especially in women for reasons we don't know yet.[10]

While it may be natural to write off one bad night's sleep as not a big deal, rarely is it ever just one bad night, and those periodic episodes of inflammation then add up to cause real harm. One of the most seminal longitudinal studies documenting the relationship between systemic inflammation and neurodegeneration was published by a large group of researchers across multiple institutions in 2017, including Johns Hopkins University, Baylor University, the University of Minnesota, and the Mayo Clinic.[11] It was based on the ongoing Atherosclerosis Risk in Communities (ARIC) study, which began in 1987, to study risk factors for atherosclerosis by following people in four communities through the years and involving more than 15,000 individuals. The 2017 study measured biological markers of inflammation in a group of 1,633 individuals whose average age was fifty-three at the start of the study. Researchers followed the participants for twenty-four years, assessing their memory and brain volumes as the years progressed. Those who originally had the highest level of inflammation in their body had an increased risk for brain shrinkage. In fact, their memory centers were 5 percent smaller in comparison to those who had lower markers of inflammation at the start. Although 5 percent may not sound like a large number, don't think of this as a linear phenomenon. Even a small percentage dip affects the ability to think and remember. In the people whose brains had shrunk, their ability to recall words was shown to be much weaker than those whose retained brain volume. These findings speak volumes and provide a cogent message for younger people who can't visualize how their habits could be affecting their long-term ability to preserve their brains. Every night of sleep counts.

MYTH: There's nothing wrong with taking sleep aids. They help you to fall asleep faster so you get more overall sleep.

TRUTH: Virtually all sleep aids, whether they are prescription or over-the-counter (OTC), will help you fall asleep faster, but they do not allow you to experience sleep as restful as natural sleep. Some even increase the risk for brain decline and dementia. Benzodiazepines (e.g., Valium, Xanax), which are often prescribed for insomnia or anxiety, are habit forming and have been

associated with developing dementia. Other sedatives, such as
Ambien and Lunesta, have been shown in clinical studies to impair thinking and balance. And common OTC drugs like anticholinergics (e.g., Benadryl, Nyquil, "PM" formulas) have been linked
to a higher chance of developing Alzheimer's. These medications have the chemical property of blocking the neurotransmitter acetylcholine, which is essential for processing memory and
learning and is decreased in both concentration and function in
patients with Alzheimer's disease. In fact, the Alzheimer's drug
donepezil (Aricept) is a cholinesterase *inhibitor*, meaning it inhibits the enzyme that breaks down acetylcholine.

THE RINSE CYCLE

Among the more recent and captivating findings about sleep has been
discovering its "washing" effects on the brain. The body clears waste and
fluid from tissues through the lymphatic system. Lymph is the colorless
fluid in specialized vessels that carries toxic waste and cellular debris.
These compounds are filtered as they pass through lymph nodes. The
lymph itself then goes back into the bloodstream. Scientists long thought
that the brain didn't have a lymphatic system and instead relied on waste
slowly diffusing from brain tissue into the cerebrospinal fluid. But then a
paper landed that rewrote the scientific narrative.

In 2012, Dr. Jeffrey J. Iliff and his team at Oregon Health Sciences
University published a description of the brain's self-cleaning function
to get rid of waste.[12] Their research ignited a new field of exploration into
the drainage pathway that is now referred to as the glymphatic system.
One year later, another paper by Dr. Iliff and two colleagues, Dr. Lulu
Xie and Dr. Maiken Nedergaard from the Department of Neurosurgery
at the University of Rochester, documented that the glymphatic system
goes into overdrive at night, suggesting that sleep provides the setting

for a cleansing or wash of sorts.[13] And failure to remove this brain trash may be linked to a higher risk of developing dementia. Just as one night of sleep deprivation can spike levels of inflammation, so can one night of bad sleep be associated with the accumulation of beta-amyloid, the brain protein that has been associated with Alzheimer's disease.[14] Moreover, there's data that now points to a relationship between higher levels of brain amyloid and incidence of depression, and this is particularly the case for those with major depressive disorder who are not responding to any treatment.[15] The University of Rochester team showed that cerebrospinal fluid flow through the brain shot up in mice only when they were sleeping.[16] This fluid, found in the brain and spinal cord, bathes and protects the central nervous system and eliminates waste products. The Rochester team hypothesized that this flow might function like the lymphatic system in the body, draining tissues of cellular breakdown products and waste for eventual disposal. Much in the way sleep tidies up our memory hub, the hippocampus, it also scrubs the brain of metabolic refuse. Sleep performs double-duty: decluttering and taking the garbage out.

Since these seminal studies, others have shown that indeed, the brain has a "clean cycle" system for washing away metabolic debris and junk, including sticky proteins that can contribute to those amyloid plaques. Dr. David Holtzman is a neurologist at Washington University's School of Medicine in St. Louis. In one of his landmark experiments, he disrupted some of the mice's sleep right when their brains would normally begin to clear beta-amyloid.[17] These sleep-deprived animals went on to develop more than two times as many amyloid plaques over about a month, compared to their well-rested counterparts. His team has also demonstrated that the difference in deeply sleeping versus wide-awake mice's levels of amyloid in the brain is about 25 percent. Over time those proteins can gather to form amyloid plaques. Think of amyloid plaques like litter in the gutter; it eventually sparks inflammation and the buildup of those tau proteins, which may destroy neurons and start the march toward Alzheimer's disease.

A vicious cycle can set in between the brain's ability to cleanse itself as it ages and the body's ability to sleep. A 2014 paper that examined how the glymphatic system works showed that the drainage rate was 40 percent worse in older mice versus young ones.[18] While we certainly cannot change some of the natural effects of aging, this information is important because sleep disturbances are common in the elderly and too often ignored or overlooked. The first goal is to figure out what might be causing the problem. Is it a medical issue such as sleep apnea or arthritis? A side effect of medications? Maybe it is a shift in the circadian rhythm that has the older person feeling sleepier earlier in the evening than when she was younger, and so she goes to bed earlier but may not sleep through the night.

Dr. Kristine Yaffe is a professor of psychiatry, neurology, and epidemiology at the University of California San Francisco, where she is director of the Center for Population Brain Health. She is world-renowned for her studies on cognitive aging and dementia and a governance member of the Global Council on Brain Health. At her memory disorders clinic, she hears a common complaint: difficulty falling asleep and staying asleep. People feel fatigued throughout the day and are compelled to nap. When Yaffe led a series of studies looking at more than thirteen hundred adults older than seventy-five over a five-year period, she documented that those with disrupted sleep had more than double the risk of developing dementia years later.[19] Most of these people had conditions that adversely impacted their sleep, such as sleep-disordered breathing, sleep apnea, breaks in their natural circadian rhythm, or chronic awakenings at night.

The other issue is the fact that Alzheimer's *itself* disrupts sleep. You can likely see the dangerous cycle here that can set in: Bad sleep prevents the brain from cleansing itself of debris, leading to extra amyloid hanging around to trigger Alzheimer's disease. The disease then sends the brain down the road to a graveyard of neurons and worsening sleep. Meanwhile, the sleep deprivation disrupts the circadian rhythm, affecting both the body's metabolism and levels of the hormone melatonin, which helps the body to sleep. This disruption in metabolism and important sleep-related hormones exacerbates the sleep disturbance, and the cycle continues. Unless this cycle is broken, the damage worsens.

What all of these studies are clearly beginning to show is the probability of a bidirectional relationship between sleep and the risk of cognitive decline. Dementia doesn't just make it hard to sleep; poor sleep may drive the development of brain decline too. More research is needed, especially among humans, but the lesson should be obvious: sleep is medicine. We need it to function during the day and refresh during the night. So with that in mind, let's turn to some strategies for a better night's sleep.

THE TOP TEN SECRETS TO SLUMBER

1. *Stick to a schedule and avoid long naps.* Get up at the same time every day, weekends and holidays included. Although many people try to shift their sleep habits on weekends to make up their sleep deficiency accumulated during the week, this can quickly sabotage a healthy circadian rhythm. If you stay up later on Friday and Saturday nights to socialize and then sleep in the next morning, you will suffer from what's called "social jet lag"; irregular sleep patterns like this are detrimental to health. The evidence on whether naps are beneficial to brain health in older adults is still unclear. If you must, limit napping to thirty minutes in the early afternoon. Longer naps later in the day can disrupt nighttime sleep. In 2019, it was reported in the journal *Alzheimer's & Dementia* that napping can be an early warning sign of Alzheimer's disease.[20] Of course, napping does not cause Alzheimer's; nevertheless, dozing off during the day could indicate that certain networks of the brain that are supposed to keep you awake are damaged. In particular, the brain areas promoting wakefulness degenerate due to the accumulation of tau, and this can happen silently early on. This may also explain why people with the disease have a tendency to nap excessively before more classic signs

of the memory-robbing illness emerge, such as forgetfulness and confusion.

2. *Don't be a night owl.* The best bedtime is when you feel most sleepy before midnight. Non-REM (rapid eye movement) sleep tends to dominate sleep cycles in the early part of the night. As night moves closer to dawn, dream-rich REM sleep begins to take over. Although both types of sleep are important and offer separate benefits, non-REM, slow-wave sleep is deeper and more restorative than REM sleep. Note that your ideal bedtime will likely change as you age. The older you get, the earlier your bedtime will become and the earlier you will naturally wake up, but your overall sleeping hours should not change.

3. *Wake up to early morning light.* Expose your eyes to sunlight first thing in the morning to help set your body clock. Everything about our evolutionary biology and neuroscience screams about the importance of mornings. We are wired to get up early and absorb the rising sun.

4. *Get moving.* Regular physical activity promotes good sleep; it can also help you achieve and maintain an ideal weight, which also can improve sleep.

5. *Watch what you eat and drink.* Avoid caffeine after lunch (definitely not after 2:00 p.m.), and don't eat or drink for three hours before bed to keep from waking up to use the bathroom. Heavy meals at dinner can also be disruptive if you consume them too close to bedtime. Be mindful of alcohol intake too. While alcohol can make you feel sleepy, its effects on the body disturb normal sleep cycles and they particularly disrupt the restorative slow-wave sleep.

6. *Mind your medicines.* Pharmaceuticals, whether over-the-counter or prescribed, can contain ingredients that affect sleep. For exam-

ple, many headache remedies contain caffeine. Some cold remedies can have stimulating decongestants (such as pseudoephedrine). Side effects in many commonly used drugs, such as antidepressants, steroids, beta-blockers, and medicines for Parkinson's, can all affect sleep too. Be aware of what you are taking, and if they are necessary medicines, see if you can take them earlier in the day when they will have the least impact on sleep.

7. *Cool, quiet, and dark.* The ideal temperature for sleeping is between 60 and 67 degrees Fahrenheit. Sleep in the dark, and minimize light sources nearby, including that from an electronic device (see tip 8). Consider a sleep mask if it is not possible to black out your environment. Try a sound machine or white noise generator to block out noises from the street if you live in an urban environment. And keep pets out of the bedroom. Bar them especially if they disrupt your sleep by moving around or making noise during the night.

8. *Eliminate electronics.* Keep the bedroom for sleeping, not looking at any type of screen, your phone included. Nearly all light—whether natural sunlight in origin or artificial from lightbulbs, TV screens, computers, and smartphones—contains blue wavelengths that are a potent suppressor of melatonin, the hormone needed for sleep, and stimulate the alert centers in the brain—a double whammy for sleep. In 2015, neuroscientist Anne-Marie Chang and colleagues showed that light-emitting devices such as e-readers caused people to take longer to fall asleep because of reduced sensations of sleepiness, decreased secretion of the sleep-inducing hormone melatonin, and a later circadian schedule, and they were less alert the next morning than people who read paper books.[21] The problem is that light-emitting diodes (LEDs) produce quite a bit of blue wavelengths, and these are ubiquitous in televisions, smartphones, tablets, and computers. Avoid blue light for a few hours before bedtime for optimum melatonin production. Use warm wavelengths in

your home LED lighting (2700–3000K is good). If you have persistent problems falling asleep, it might be easier to get eyeglasses that filter out blue light. Make sure that your clocks, nightlights, dimmers, and so on use red, or "warm glow," lights rather than blue or green. Red light has the least power to shift circadian rhythm and suppress melatonin. Get an app that changes the color temperature of your screen to avoid blue light, particularly if you like to read in bed.

9. *Establish bedtime rituals.* Try to set aside at least thirty minutes to an hour before bedtime to unwind and perform tasks that help your body know that bedtime is coming soon. Disconnect from stimulating tasks (such as working, being on the computer, or using a cell phone) and engage in activities that are calming, such as taking a warm bath, reading, drinking herbal tea, or listening to soothing music. Stretch or do something relaxing. Wearing socks to keep your feet warm can also help you fall asleep more easily. Stay away from difficult conversations, and keep everything peaceful before bedtime. No arguing or discussing touchy, contentious topics either (issues always seem better in the morning anyhow).

10. *Know the warning signs.* You may have a bona fide sleep disorder that could benefit from treatment if you have several of the following symptoms: trouble falling or staying asleep three times a week for at least three months; frequent snoring; persistent daytime sleepiness; leg discomfort before sleep; acting out your dreams during sleep; and grinding your teeth or waking with a headache or aching jaws.

If you have tried all of these approaches and still fail to get a good night's sleep or find yourself relying on sleep aids regularly, talk to your doctor about your sleep. He or she might recommend that you undergo a sleep study to rule out issues such as undiagnosed sleep apnea. This will require that you spend the night in a sleep lab that can monitor and record your sleep. These centers are not as unusual as you might think. Many hospitals large and small offer these services.

DON'T FORGET DAYTIME R&R

Sleep is a rejuvenating activity that the body demands, but there is a difference between sleep and rest. We need both and must include other activities of rest and relaxation into our waking lives in order to stay sharp. Our mental well-being in general depends on this too—and we know that greater mental well-being is associated with reduced dementia risks. We also know the opposite is true: Conditions like certain types of anxiety and depression can be warning signs for cognitive decline and Alzheimer's disease. So combating these by reducing stress and building mental resilience is important.

I'm a big proponent of meditation. I practice it every day using a type known as analytical meditation. I picked up the habit after spending time with the Dalai Lama a few years ago at the Drepung Monastery in Mundgod, India. I'll admit that at first, I wasn't sold on the idea. I was terrified! Just thinking about meditating with His Holiness was making me anxious. But who says no to a chance to meditate with the Dalai Lama? I agreed to join him early one morning at his private residence.[22]

All my meditation insecurities immediately started to kick in once I was sitting with him cross-legged and trying to focus on my breathing with my eyes closed. After a few minutes, I heard his deep, distinctive baritone voice: "Any questions?"

I looked up and saw his smiling face, starting to break into his characteristic head-bobbing laugh.

"This is hard for me," I said.

"Me too!" he exclaimed. "After doing daily for sixty years, it is still hard."

It was at once surprising and reassuring to hear him say this. The Dalai

Lama, Buddhist monk and spiritual leader of Tibet, can also have trouble meditating.

"I think you will like a type of analytical meditation," he told me. Instead of focusing on a chosen object, as in single-point meditation, he suggested I think about a problem I was trying to solve, a topic I may have read about recently, or one of the philosophical areas from our previous discussions. He wanted me to separate the problem or issue from everything else by placing it in a large, clear bubble. With my eyes closed, I thought of something nagging at me—something I couldn't quite solve. As I placed the physical embodiment of this problem into the bubble, several things started to happen very naturally.

The problem was now directly in front of me, floating weightlessly. In my mind, I could rotate it, spin it, or flip it upside down. It was an exercise to develop hyperfocus. As the bubble was rising, it was also disentangling itself from other attachments, such as subjective emotional considerations. I could visualize it as the problem isolated itself and came into a clear view.

Too often, we allow unrelated emotional factors to blur the elegant and practical solutions right in front of us. It can be dispiriting and frustrating. Through analytical meditation, His Holiness told me, we can use logic and reason to more clearly identify the question at hand, separate it from irrelevant considerations, erase doubt, and brightly illuminate the answers. It was simple and sensible. Most important for me is that it worked.

As a neuroscientist, I never expected that a Buddhist monk, even the Dalai Lama, would teach me how to better incorporate deduction and critical thinking in my life, but that is what happened. It changed me, and I am better for it. I practice analytical meditation every day. The first two minutes, as I create my thought bubble and let it float above me, are still the hardest. After that, I reach what can best be described as that quintessential *flow* state, in which twenty to thirty minutes pass easily. I am more convinced than ever that even the most ardent skeptics could find success with analytical meditation.

Over the holidays, I spent as much time as possible relaying the Dalai

Lama's teachings to my family and friends and teaching them basic principles of analytical meditation. This was the gift I most wanted to share with them—and now with you. It is an ingredient to get crucial rest in your day, one that is different than sleep.

Practices in mindfulness are on the rise. In 2018 the Centers for Disease Control and Prevention released a report stating that between 2012 and 2017, participation in yoga increased 50 percent, from 9.5 to 14.3 percent, and the use of meditation more than tripled from 4.1 to 14.2 percent.[23] These practices share a common theme: being present in the moment and observing what's happening in your life. We often hear anecdotally that mindfulness activities can combat stress, but it's important to know that this idea has been well substantiated in the medical literature. These habits are even making their way into places where you'd least expect them: military combat zones. In 2014, for instance, a group of Marines received training in mindfulness-based techniques and were subsequently found to have enhanced cardiovascular and pulmonary recovery after exposure to high-stress simulations of military activity.[24]

You don't have to be a soldier to benefit from this training. One effect that is easily generalized to all of us is that mindfulness practices also appear to lower levels of the stress hormone cortisol. In one of the most comprehensive and well-cited studies in this area, a meta-analysis published in the *Journal of the American Medical Association* reviewed all relevant trials on the subject, finding that mindfulness significantly reduces anxiety, depression, and pain.[25] Another meta-analysis looked at the effect of transcendental meditation, a type of mindful practice that involves the use of a mantra, on 1,295 people across sixteen studies.[26] It too found that the practice led to significant diminishment in anxiety, which became even more pronounced in those starting off with high levels of it.

Meditation has a long and storied history that has only recently gained scientific validity. Researchers are finally beginning to understand how it can affect the aging process itself. It started in 2005 when Harvard's Massachusetts General Hospital published an imaging study showing that particular areas of the cerebral cortex, including the prefrontal area, were thicker in people who frequently meditated.[27] Since then, numerous follow-up studies by the same group and others around the world have documented that "thick-brained" people tend to be smarter and have stronger memories. These cortical areas help with attention and sensory processing and are used for planning complicated cognitive actions.

The so-called relaxation response that's achieved through meditation can also happen through various forms of yoga, tai chi, breathing exercises, progressive muscle relaxation, guided imagery, and even repetitive prayer. One of the reasons that deep breathing, for instance, is so effective is that it triggers a parasympathetic nerve response, as opposed to a sympathetic nerve response, the latter of which is sensitive to stress and anxiety. When you perceive stress, the sympathetic nervous system springs into action, resulting in surges of the stress hormones cortisol and adrenaline. The parasympathetic nervous system instead can trigger a relaxation response, and deep breathing is one of the quickest means of getting there. In a deeply relaxed state, your heartbeat calms down, breathing slows, and blood pressure lowers.

Deep breathing can be done anywhere, anytime. If you've never meditated before, a deep breathing practice twice daily will get you started and give you a foundation for trying more advanced techniques. All you have to do is sit comfortably in a chair or on the floor, close your eyes, and make sure your body is relaxed—releasing all tension in your neck, arms, legs, and back. Inhale through your nose for as long as you can, feeling your diaphragm and abdomen rise as your stomach moves outward. Take in a little more air when you think you've reached the top of your lungs. Slowly exhale to a count of twenty, pushing every breath of air from your lungs. Continue for at least five rounds of deep breaths.

Mindfulness can be achieved in all kinds of ways, from using an app on your phone to guide you through a fifteen-minute practice, to participating in a restorative yoga class, to Japanese forest bathing, or *shinrin-yoku*—which means just being in the presence of trees. Forest bathing has been popular lately as a way to lower heart rate and blood pressure and reduce stress hormone production. When you are forest bathing and breathing in the "aroma of the forest," you are also absorbing substances known as phytoncides, which protect trees from insects and other stressors. As we have learned over the past decade, these phytoncides can also protect us by increasing our natural killer immune cells and decreasing cortisol levels.[28] While spending time in nature or green spaces has long been recommended to improve mental well-being, we now understand what that aroma of the forest is really doing for our bodies and brains. You needn't travel to a far-off forest; you can do well by yourself just by digging in the dirt of your own garden or visiting a local park. I have always loved the ancient Indian concept of creating a harmonious 100-year life by spending the third stage (around the ages fifty to seventy-five) living in a forest as part of a contemplative, tranquil lifestyle called *vanprastha* (life as a forest dweller). Some research has found that walking in nature, as opposed to walking in urban environments, may help people manage stress, calm rumination, and regulate emotion.[29] A number of studies have found that green spaces and parks in cities are linked to positive mental health.[30] I spend a lot of time indoors—and in windowless operating rooms—so I cherish the times I can roam and play outside and absorb the pleasures of nature.

Here are some other R&R ideas to consider in the name of mental well-being. These strategies also have the effect of helping you build a more resilient, productive brain:

Become a regular volunteer in your community. Those who volunteer tend to have less anxiety, depression, loneliness, and social isolation, as well as a sense of purpose. The 2018 AARP survey found that adults age fifty or older who volunteer at least once a year have higher mental well-being scores than those who don't volunteer at least once a year.[31] Consider taking a leadership role in a group or organization of which you are a part.

Express gratitude. Begin or end your day by thinking of things for which you are grateful. Consider keeping a gratitude journal.

Research finds that gratitude reduces depression and anxiety, lowers stress, and increases happiness and empathy.[32] It is hard to be angry or distressed when you are practicing gratitude. My active gratitude practice is a big part of giving my brain a time-out. It acts like a big reset button on my brain and lets the less significant issues (that are a disproportionate brain drain) melt away. I do this myself and with my family every day I can.

Practice the art of forgiveness. Positive psychology research has found that forgiving oneself and others promotes life satisfaction and self-esteem.[33]

Look for things that make you laugh. Explore humorous movies, books, or online videos. Laughter triggers the secretion of "feel good" hormones such as endorphins, dopamine, and serotonin, which can relieve stress and reduce tension and anxiety—even lessen pain.

Take breaks from email and social media. Consider turning off notifications. Put your smartphone in another room, and turn off the sound to help you focus on a task. Consider checking social media (e.g., Facebook, Instagram) on a schedule and avoiding smartphones during meals and family time. Avoid emails first thing in the morning. Mornings are golden time. Use them to do your most creative work as opposed to procedural.

Find another hour in your day at least once a week. If you want to "create" another hour in your day, be very strict with your time spent looking at a screen (computer, phone, TV, tablet). If you set aside one day a week to be screen-free, I bet you'll find at least an hour extra of time to do whatever you like.

Establish a system of rewards. The brain and body like rewards; anticipating a reward can release a little dopamine surge. The Pomodoro technique works for this reason. It's a proven strategy in gaining the most out of your time by using mini-breaks at timed intervals as rewards. And it's easy: Simply pick a task—hopefully your most important one of the day—and set your timer for twenty-five minutes. Work solely on that task with no distractions until the timer rings. Then take a five-minute break and repeat as necessary.

Don't multitask—tackle your day like a surgeon. Contrary to our attempts to manage multiple tasks at the same time, the brain doesn't like to do that. Surely you can walk and talk at the same time while digesting your lunch, but the brain can't concentrate on executing two activities that demand conscious effort, thinking, comprehension, or skill. Would you want me operating on your brain while I write an email and take a phone call? The brain handles tasks sequentially but can switch *attention* between tasks so quickly that we're given the illusion that we can perform multiple tasks together. So if you want to get more done using less effort, aim to work on what's called your attentional ability: focus and concentrate on one sequence—one task—at a time and avoid distractions. This can be a surprisingly joyous experience—one that I experience whenever I am in the operating room. The OR is one of the few places where distractions aren't allowed. You are scrubbed, unable to check your phone as you enter a fully focused state on the task in front of you. It is like taking your turbo-powered brain on an empty flat road and letting it rip. Most of the time, our brains are stuck in stop-and-go traffic, working hard and getting nowhere. Let your brain roam free every now and then. You will not only get more done than you thought possible, but you will also achieve a level of bliss that is tough to replicate. When you try to mentally multitask, you slow down your thinking and everything takes longer to accomplish. The brain loves the rhythm of sequences. It also helps with sanity!

Identify your marbles and sand and plan accordingly: If you have a jar that you are filling with marbles and sand, which do you put in first? The marbles. Then you can allow the sand to fill the spaces in between. This is a key metaphor for planning your day and maximizing your time. Think of the marbles as the important blocks of your day (appointments, commitments, projects, important tasks including exercise and sleep), and the sand is everything else (checking email, returning a call, dealing with nonurgent things). Don't get stuck in the sand. Tip: Plan to set aside thirty minutes every Sunday night for your weekly check-in and ask yourself this powerful question: "What goals do I need to accomplish in the next seven days for me to feel this week was a success?"

Declutter your life. Clean out closets, basements, storage places, and garages. Donate old clothes and books that no longer bring you pleasure. Toss old magazines and catalogs. Throw away or shred mail, bills, and letters that you don't need. Make a habit of immediately throwing away things that don't have relevance personally or create value productively. Overall, manage your environment. Mess creates stress, as disorganization equates with distraction.

Set aside fifteen minutes each day for yourself. Use this time to engage in a de-stressing activity such as meditation, which can be as simple as sitting quietly for a few minutes and focusing on taking deep, calming breaths. There are smartphone apps and web pages for guided meditation that can help. Or use this time to write in a journal. Avoid anything too stimulating or distracting like scrolling through social media or shopping online. The key is to get to really know yourself, and most of us are not that good at self-awareness. I wasn't taught that in medical school, but it has since become a vital part of my personal compass. We are all different, and the best person to help guide you is *you.*

Let yourself daydream. The mind can't stay in one gear all day long. We tend to force the brain to be directing our thoughts as much as possible, rather than let the thoughts take over and run themselves. Daydreaming, however, can act as a neural reset button.

Do not be afraid to seek help from a health professional if you have concerns about your mental health. Conditions such as anxiety and depression are common and can be treated.

LIFE'S TRANSITIONS

It's important to acknowledge that we all go through phases in life that bring different challenges. With age come transitions punctuated by events like the birth of children, the death of loved ones, shifts in marital status, changes in financial means, retirement, accidents, illness, perhaps the loss of some independence such as the ability to drive. People who can adapt to changing life circumstances and experiences may more quickly resume close-to-normal feelings and states of mental well-being. Prolonged sadness or bereavement is not part of the normal response to these transitions and ups the risk for cognitive impairment.

There is a surprising silver lining with age, however. Despite feelings of loss that often occur as people age, getting older does not necessarily mean people become less happy. On average, most people report greater mental well-being as they age past their mid-fifties into the later stages of life. This tendency to report high levels of happiness and well-being around the ages of eighteen to twenty-one, declining through young adulthood and midlife with significant increases starting around age fifty, is often described as a U-shaped curve of happiness over the life span.[34] People tend to be happier when they are younger and older but have a decrease in happiness during midlife. A number of studies have also found a positivity effect of aging, meaning that older adults tend to remember and pay attention to positive information more than negative information.[35]

Why is midlife—ages thirty-five to roughly fifty-five—such a downer? It's typically the period when stressors peak: You are straddling the competing demands of aging parents and dependent children, all while working to maintain a career and save for retirement. This U-bend of life does have its critics, and it can be hard to generalize happiness across different populations in their lifetime, but I bring it up because it's often part of the conversation.

It's important that you do what you can to keep track of your mental well-being and seek help when your stress levels reach toxic levels. Although scientists don't think that depression, especially at midlife, causes dementia later in life, this area continues to be studied. Depression is a risk factor for dementia, but we don't know if the relationship is cause-and-effect or just correlated somehow. We do have evidence that people who experience depression in later life (over the age of fifty) are more than twice as likely to develop vascular dementia and 65 percent more likely to develop Alzheimer's disease than people who are not depressed.[36] And people who do develop dementia and have a history of depression often have an increase in new symptoms of the problem about a decade before their dementia becomes evident.

In 2019, my HBO special, *One Nation under Stress*, aired after I'd spent two years traveling around the country in search of understanding why "deaths of despair" due to suicide and overdoses are on the rise. Too many people suffer from what I call toxic stress that fuels levels of unimaginable depression. The goal of my film was to call attention to society's need for better ways to cope with life's ups and downs. Since then, it's been reassuring to see more attention given to mental health issues, though it became clear to me as I traveled all over the country that we all need to do a better job taking care of each other.

Sleep and daytime stress-reducing activities can work wonders on the brain and body, but they are not the only habits we must maintain to keep sharp and mentally well. As you're about to find out, there's a tremendous power in what we put on our plates.

CHAPTER 7

Food for Thought

The only way to keep your health is to eat what you don't want,
drink what you don't like, and do what you'd rather not.

MARK TWAIN

Mark Twain's humor is timeless. His quip about health may still be partly true today, more than a century later. But Twain's witty observation carries a hidden truth: knowing what to eat for good health can be bewildering even by modern standards. I am amused by the number of diet books and related material that come out each year, often on cue around the "new year, new you" campaigns. Yet there is endless confusion about the ideal way to fuel your body—whether the goal is ("effortless") weight loss, preventing heart disease, boosting brain function, or something else entirely.

Think about your own experience. How many times have you wondered: Paleo, keto, gluten free, low carb, low cholesterol, pescatarian, low fat, vegan? Those are just a small slice of the diets that have been reported over the past couple of years. Doctors seldom discuss nutrition with their patients. Once again, consider your own experience. When was the last time your doctor spent time with you going over your dietary choices and making science-backed suggestions? In a 2017 article published in *JAMA*, Dr. Scott Kahan of Johns Hopkins and Dr. JoAnn Manson of Harvard address the problem of this important subject getting left out of the conversation during office visits.[1] The consequence, they say, is that "patients receive most of their nutrition information from other, and often unreliable, sources." They cite that only 12 percent of office visits include counseling about diet. So if your doctor has had this talk with you (and you were honest in your replies to questions), consider yourself lucky.

About once a year, a compelling new protocol gains the spotlight, often backed by a weak or dubious hypothesis and cherry-picked science—what's called "data dredging." This helps explain why we read so many contradictory nutrition headlines. Today red wine, coffee, and cheese are found to be protective against dementia (and heart disease and cancer), and tomorrow another pair of studies pronounce that they do the opposite. Which brings me to the question I set out to answer: What *is* the best possible diet for my brain? Does it exist? *Can* it exist? Would Mark Twain wish he'd lived in the twenty-first century?

To get at the core of this issue, I spent countless hours with experts all over the country and synthesized a great deal of information because there is no consensus on the answer to this question. Nailing down any sort of conclusion is like trying to throw a blunt dart with a bad arm and hit a moving target. I was in fact astounded at how contentious the debate about diet can be (and the comical fact that the word *die* is within this four-letter word). Many of the foremost experts on the brain disagree on answers to basic questions that I thought would be straightforward and uncontested. Is gluten harmful to the brain? Are ketogenic diets hyped? Do "superfoods" for the brain truly exist (and just what qualifies as a "superfood")? Is there a time and place for supplements and vitamins to make up for dietary flaws? As the late Senator Daniel Patrick Moynihan put it, "Everyone is entitled to his own opinion, but not to his own facts." That statement couldn't ring truer when it comes to the plate debate. The problem, however, is we don't have all the facts. The experts themselves can't even agree on the difference between opinion and fact.

For starters, I am comfortable saying this: We do have evidence that how you fuel your body can go a long way toward protecting your brain. That conclusion sounds simple but has been based on decades of research that has finally borne fruit. Dr. Manson has stated, "I've been impressed by the compelling evidence that nutrition and lifestyle drive down the risk of the major chronic diseases in the United States—type 2 diabetes, cardiovascular disease, cancer, and down the line [think dementia]. The evidence for this has reached a critical mass."[2] Her passion for getting the message out encouraged her to move away from a focus on clinical practice to one on population health and prevention research in an effort to address the risk factors of chronic disease rather than just disease management.

You can breathe a sigh of relief: I'm not talking about following a particular branded diet. I'm referring to a *way of eating*—a dietary style with general guidelines. This seems to make the biggest difference in health both short term and long term. When Sara Seidelmann, a cardiologist and nutrition researcher at Brigham and Women's Hospital in Boston, looked at the eating habits of more than 447,000 people around the world, she found that no matter where you live or what your daily diet is like, avoiding entire food groups or restricting certain foods under the thinking you can game your way into good health is not an ideal approach. It might work for a while, but it could also backfire and hasten your death. Her advice, published in 2018 in the *Lancet*, echoes some unsexy, old advice: everything in moderation.[3] I will add another reminder as well: We are all different, and getting to an ideal way of eating for you may be a little (or a lot) different than it is for someone else. Part of the solution is figuring out what really fuels you in the best way without digestive issues or food allergies. If you focus more on what you should eat instead of what you shouldn't eat, you will end up fueling with good calories and naturally avoid the bad ones.

Say good-bye to following strict dietary protocols that are unrealistic and challenge your willpower. I titled this chapter "Food for Thought" for good reason: You'll be given a general framework for creating meals that satisfy your preferences while staying on a path that fosters brain health. Besides, if you worry too much about "eating right," you'll raise your anxiety and increase cortisol levels, which would be more dangerous than the benefits of the "right diet" for brain health! Food should be a source of nutrition, yes, but it should also be a source of enjoyment. I go out of my dietary lane from time to time and have no guilt when I do. Guilt is bad for the brain, and too much of it makes you lose your sharpness.

What makes this area of medicine so tricky and controversial is that nutrition studies generally are limited. It's very difficult, if not impossible, to conduct traditional studies on diets using a randomized, controlled design. These investigations cannot be compared to pharmaceutical studies because we cannot use a true placebo group to study essential nutrients. We can't deprive people of certain nutrients they need to live just for the purposes of conducting a study. Also bear in mind that foods contain a staggering number of different biomolecules. If we find associations between a particular type of food and a health effect, the exact molecules that produce the desired effect are difficult or impossible to isolate because of the complex composition of foods and potential interactions among nutrients. Plus, there are underlying genetic factors to consider in consumers themselves. There's also the practical issue of basing a nutritional study on people's honest recollections of what they ate (Do you remember what you had for dinner last Tuesday? Will you confess to that decadent chocolate dessert last night?), as well as controlling for lifestyle (How many times did you break a sweat last week? Did you smoke any cigarettes? How many?). All of these variables and more can factor into the diet equation.

In 2018, these complexities triggered the retraction of a seminal study published in 2013 in the prestigious *New England Journal of Medicine* that endorsed a Mediterranean-style diet, which you've no doubt heard about over the years as being beneficial. Among the first studies to shine a positive light on Mediterranean-style diets, which are rich in olive oil, nuts, plant proteins, fish, whole grains, fruits and vegetables, and even wine with meals, was the PERIMED (Prevención con Dieta Mediterránea) research project conducted in Spain in the mid-2000s and published in the *Annals of Internal Medicine*.[4] It concluded that such a dietary style could lower cardiovascular risk factors. The 2013 study showed people aged fifty-five to eighty who ate a Mediterranean diet were at lower risk of heart disease and stroke—by as much as 30 percent—than those on a typical low-fat diet. In 2018, the authors of the 2013 study published a reanalysis of their data in the same journal following criticism about their methodology.[5] Although there were flaws in their original study, mainly due to the limitations of controlling for factors I already mentioned, their overall conclusion remained the same. Plenty of other studies have also shown that people who adhere to a Mediterranean-style diet enjoy greater brain volume as they age compared to their counterparts who don't eat that way.

Dr. Martha Clare Morris, a professor of epidemiology at Rush University in Chicago and director of the Rush Institute for Healthy Aging, was a founding member of the Global Council on Brain Health. Prior to her death in 2020, she conducted pioneering work to find effective dietary protocols to prevent Alzheimer's disease. In 2015, she published the MIND diet for healthy brain aging, based on years of research into nutrition, aging, and Alzheimer's disease.[6] It was followed by her book *Diet for the MIND*.[7] Her research focuses on studies that live up to the scientific method as much as possible despite the inherent limitations of nutrition studies. When I spoke with her in 2018 about her study, she was excited that her investigations were among the first to show the effects of diet on the brain. Although she acknowledged the limitations of nutrition studies, she believed we are finally able to make data-driven suggestions about what we should be eating.

The MIND diet was created by taking the basics of two popular diets—the Mediterranean and DASH (Dietary Approaches to Stop Hypertension)—and modifying them to incorporate science-supported dietary changes that improve brain health. MIND is a catchy abbreviation; it stands for Mediterranean–DASH Intervention for Neurodegenerative Delay. And there's nothing surprising about the diet: thumbs up for vegetables (especially green leafy ones), nuts, berries, beans, whole grains, fish, poultry, olive oil, and, for those interested, wine; thumbs down on red meats, butter and stick margarine, cheese, pastries and other sweets, and fried or fast food. What might surprise you is how well this diet works. In her reasonably well-controlled study on this diet over ten years of nearly a thousand people, she showed it could measurably prevent cognitive decline and reduce the risk of Alzheimer's disease. People who had the lowest third of MIND diet scores (meaning they followed the diet less) had the fastest rate of cognitive decline. People who had the highest third of scores had the slowest rate of decline. The difference between the highest third and lowest third in cognitive decline was equivalent to about seven and a half years of aging. I'll take back seven and a half years of aging, and I am sure you would as well. People who were in the highest third of MIND diet scores had a 53 percent reduction in the risk of developing Alzheimer's, and those who had the middle third of scores for following the MIND diet still enjoyed a 35 percent reduction in the risk of developing the disease.

So despite the challenges in conducting nutrition studies, we do have data to show the direct impact of nutrition on the brain and we are arriving at the best ways to feed it. We have enough evidence between outcomes in human clinical trials, mouse models, and epidemiological studies to make certain assertions with confidence. And I know that deep down you already knew that eating muffins every morning for breakfast with a mochaccino probably wasn't going to get you where you really needed to go. Diets may seem confusing, but food isn't.

MYTH: Superfoods like kale, spinach, nuts, and seeds will protect your brain.

TRUTH: The term *superfood* has no medical meaning whatsoever. Although it implies that a food provides health benefits, it's a marketing term the food industry uses to sell more product. Some foods with the superfood halo on them can be supergood for you, such as fresh blueberries and a handful of omega-3-rich macadamia nuts, but be careful about claims that they do something specific for the brain. And there are "superfoods" sold out there that are anything but; juice drinks made with 100 percent fruits are mostly sugar and are stripped of what made those fruits super to begin with, fiber.

WHAT'S GOOD FOR THE HEART IS GOOD FOR THE BRAIN

Over the course of my career, I've witnessed a sea change in how we view the relationship between diet and brain health. Once the science spoke and doctors listened, the mantra became, "What's good for the heart is good for the brain." That statement doesn't paint the whole picture, but it's not a bad place to start. Common conditions influenced by diet such as elevated blood pressure, high cholesterol, and diabetes harm both cardiovascular and cognitive health. Because you are reading this book, you probably already know that, especially if you suffer from any of these conditions. But separately, and more precisely, we can also say that a heart-healthy diet is a brain-healthy diet.

Recent studies evaluating the incidence of dementia among large groups of people over several decades have found decreases in dementia occurring simultaneously with improvement in cardiovascular health. The 2017 AARP Brain Health and Nutrition survey, released in early 2018, also found that significantly more adults age fifty and over without heart disease rated their brain health/mental sharpness as "excellent" or "very good" compared to those with heart disease.[8] The connection between the heart and brain goes far beyond the fact that the brain receives blood from the heart. It is important to remember, however, that the brain functions uniquely, and often separately from the rest of the body. There is even a barrier—the blood-brain barrier—that acts like a gated door: only certain molecules crucial to neural function are allowed into the brain from the blood. This is what makes the brain independent to some degree.

My search for more insights into diet and brain health specifically took me to neurologist Dr. Richard Isaacson, the director of the Alzheimer's Prevention Clinic at Weill Cornell—a groundbreaking prevention clinic that's on the cutting edge of medicine in the field of brain health. He is also the coauthor of *The Alzheimer's Prevention and Treatment Diet*.[9] Initially, the dean of the medical school thought Isaacson was crazy to establish a "prevention" clinic because Alzheimer's disease had always been considered unpreventable. But times—and the thinking—have changed. Clinical trials are cropping up around the world now to study lifestyle interventions that have protective effects in people who are at an increased risk for cognitive decline and dementia. One, the Finnish Geriatric Intervention Study to Prevent Cognitive Impairment and Disability, or FINGER Study, led by Dr. Miia Kivipelto, also a founding governance member of the Global Council on Brain Health, was completed in 2014 and reported that a two-year combination therapy that targeted things like a healthy diet and exercise found that these strategies can indeed help preserve cognition. In the United States, the Alzheimer's Association is heading the U.S. Study to Protect Brain Health Through Lifestyle Intervention to Reduce Risk (U.S. POINTER), which also involves a two-year clinical trial. And in New York, Dr. Isaacson is making his own splash in these previously uncharted waters.

The Cornell dean took a gamble on Dr. Isaacson, impressed by his credentials at such a young age (he was barely thirty when he pleaded his case for his clinic), and was willing to let him "do the screening thing." Now Dr. Isaacson oversees teams of people who build tech applications, assist with his research programs, and develop new methods of cognitive testing. In late 2018, his work made the cover of *Alzheimer's & Dementia*, one of his field's most prestigious journals and the flagship journal of the Alzheimer's Association.[10] The following year, his seminal study was presented at the annual conference for the Alzheimer's Association and published in the same journal.[11] This study made mainstream headlines for good reason: he has demonstrated that people can delay the progression of cognitive decline due to the aging process by *two to three years*, on average, through simple lifestyle interventions and even if they have a family history of Alzheimer's disease. "Alzheimer's disease begins in the brain decades before the first symptoms of memory loss, leaving ample time for people at risk to make brain healthier choices," he reiterated to me. "Our study showed that people can be proactive and work together with their doctors to not only improve cognitive function, but also reduce their Alzheimer's and cardiovascular risk. On average, people received twenty-one different recommendations that were personalized for them. Considering the results of this study, along with the totality of prior evidence, people should feel empowered to take control of their brain health starting today. One in three cases of Alzheimer's may be preventable if that person does everything right, and I believe that individualized management is our most promising way forward in the fight against Alzheimer's disease." His methods are stirring a revolution in brain medicine. Unlike his predecessors, who didn't consider dietary effects on the brain, Dr. Isaacson "prescribes" certain foods to his patients because he knows that nutrition matters. And he sees a difference in patient outcomes. He also prescribes other basic lifestyle strategies such as exercise, sleep, and stress management—which I will describe in detail at the end of Part 2. I truly believe he's setting a new model for tackling brain health and disease in the twenty-first century. People who were diagnosed with MCI at the start of this study and who followed at least 60 percent of the recommendations showed cognitive improvement.

Dr. Isaacson takes a novel approach to traditional ways of managing the disease and likens his methods to how we prevent and treat other chronic diseases such as hypertension and diabetes. Preventing—and treating—dementia requires a personalized plan for each individual because no single patient is the same. While patients may look similar in symptoms and pathology, the forces driving their disease and their individual risk factors can be very different, so what works for one person may not help the next. His philosophy is in line with the tenets of what future medicine will be for all of us: precision medicine whereby we are given specific, comprehensive protocols and prescriptions tailored to our physiology and needs. Customized care takes into consideration our genes, environment, and lifestyle. Isaacson likes to focus his efforts on prevention because he knows the disease starts decades before there are any outward signs. To support his mission, he has launched free online courses at AlzU.com that laypeople (and doctors) can take to educate themselves about brain health and learn about ongoing research that's translated for a general audience. In part 3, I share more details about his astonishing results from his interventional studies. Dr. Isaacson is among the first scientists to document the beneficial effects of lifestyle habits on the risk for cognitive decline and lessening symptoms. Best of all, he's revealing improvements in as little as eighteen months after he puts patients on his programs—some of whom are in their twenties with no obvious signs of cognitive problems but who want to stack the deck in their favor so they may be able to avoid dementia entirely as they get older.

Dr. Isaacson has focused much of his practice on reducing risk (both he and his brother were inspired by their family's health history to become neurologists). The experience that really struck Isaacson involves his Uncle Bob. When Isaacson was three years old, he fell into his aunt's pool and sank to the bottom. Uncle Bob, who was in the navy at the time, jumped in and rescued him. When Isaacson was in high school and applying to medical programs, Bob was diagnosed with Alzheimer's disease at age seventy. Isaacson was crushed and kept wondering if he could develop a treatment to help the man who had once saved his life. His mission in life was set.

Dean Ornish's mission is not too different. At his Preventive Medicine Research Institute in the San Francisco Bay Area, he and his colleagues, including Dr. Bruce L. Miller, director of the UCSF Memory and Aging Center, are conducting randomized, controlled clinical trials to determine if the progression of early to moderate Alzheimer's disease can be reversed by a comprehensive lifestyle medicine program—without drugs, devices, or surgery. At the heart of his protocols is diet, among other basic (noninvasive, inexpensive) changes anyone can make. Dr. Ornish has long been a proponent of dietary interventions to treat, and sometimes reverse, a wide array of chronic diseases such as coronary heart disease, type 2 diabetes, early-stage prostate cancer, high blood pressure, elevated cholesterol levels, and obesity. The author of multiple best-selling books, including his latest *UnDo It!*, he has been a pioneer in the field of lifestyle medicine and has now set his sights on tackling Alzheimer's disease.[12] He believes we are at a stage of scientific evidence very similar to where we were forty years ago regarding coronary heart disease. In other words, epidemiological data, anecdotal clinical evidence, and animal studies show that Alzheimer's disease may be prevented or slowed by making comprehensive lifestyle changes.

The whole idea of preventing Alzheimer's or even mitigating symptoms after a diagnosis is a twenty-first-century concept. After hearing from researchers all over the world, I believe such an achievement is within our grasp, and it likely starts with how we fuel our bodies. What you eat could very well be one of the most influential benefits to your brain health now and in the future. After all, you eat every day and how your body responds to what you put in your mouth ultimately influences your entire physiology—all the way up to your brain.

While no single food is the key to good brain health, a combination of healthy foods will help secure the brain against assault, and it is never too early to begin. Think about it. The food you eat in your youth can start to lay the groundwork for protecting your brain in your later years.

It should come as no surprise that the typical Western diet—high in salt, sugar, excess calories, and saturated fats—is not brain-friendly. As the research concludes, a plant-based diet that is rich in a variety of fresh whole fruits and vegetables, particularly berries and green leafy vegetables, is associated with better brain health. I know you have heard that countless times, and you may be getting a little numb to it. Me too. But there are a few simple statistics I often share with my patients to make the point, like this one: "Increasing fruit intake by just one serving a day has the estimated potential to reduce your risk of dying from a cardiovascular event by 8 percent, the equivalent of 60,000 fewer deaths annually in the United States and 1.6 million deaths globally."[13]

The good news is that exceedingly small shifts can have a tremendously outsized effect. Who can complain about reaching for a juicy apple or a sweet handful of blueberries? Remember, we're talking about a style of eating, not a rigid *eat this, not that* strong-arming directive. Only 10 percent of Americans get the recommended number of fruits and vegetables a day. In 2018, it was reported that more than a third of us eat fast food daily.[14] At least one meal a day comes from a pizza box or a drive-through. And here was a surprise: fast food intake rises with income.

But eating well means eating real food—not popping pills and supplements. While we all like the idea of a pill with the micronutrients neatly packaged in one swallow, that approach is not effective and not really possible. That bottle with broccoli on the label doesn't really have broccoli in a pill. The evidence shows that micronutrients such as vitamins and minerals offer the greatest benefit when consumed as part of a balanced diet because all those other components in healthy food allow the micronutrients to be well absorbed and do their job better. Think of this as an "entourage effect." While there may be some star players, they don't work as well without the entourage of other ingredients. In other words, getting your B vitamins from eggs and your omega-3 fatty acids from fish trumps taking vitamins and supplements alone.

Changing your diet in an effort to optimize your brain will take some time, I realize—and it should. Most of us have a general idea of what's good for us, and what we like and don't like. I kept a food journal a few years ago to figure out what worked best for me. Fermented foods like pickles are a secret weapon for me but maybe not for you. I occasionally snack on them to boost my productivity. Find what works for you and make it part of your routine. In chapter 9, I offer food plan ideas so you know how to build in the right kinds of foods throughout your day and individualize the plan for you. One thought I share now is to aim for seven different-colored foods (real food—not jelly beans) every day. This usually has the impact of giving you all you need in terms of macro and micro nutrients. It may be a little harder than you think. Quick: Can you name seven different-colored foods?

Over the past several years, I have focused on creating a style of eating that I can easily maintain even when I'm on the road, but it does require planning and commitment. You should strive to do the same, which might require learning new methods for grocery shopping and finding the best, freshest foods for you and your family that meets your budget. What you should do immediately, however, is stop the external attack on your brain. Reducing your intake of sugar and artificially-sweetened beverages, fast food meals, processed meats, highly salty foods, and sweets is no longer a gentle suggestion; it is a mandate. Stop buying foods that a gardener or farmer (or your great-grandmother) wouldn't recognize. When you replace potato chips and processed cheese dip with nuts or carrots and hummus, you lower trans fats and saturated fats while still having a satisfying snack. This is an easy one and it turns out to be incredibly helpful to your brain.

According to that same 2017 AARP Brain Health and Nutrition survey, adults age fifty and older who get the recommended amount of fruits and vegetables in a typical day report significantly better brain health compared to those who do not get the recommended amount (70 versus 61 percent).[15] The survey found that the more fruits and vegetables men and women consume, the more likely they are to rate their brain health higher. Of those who said they don't eat any vegetables, fewer than half (49 percent) considered their brain health as "excellent" or "very good."

MY GUIDE TO GOOD EATING

With the diversity in cultural practices and lifestyle habits around the world, there are many ways to approach dietary choices. I know my three daughters eat differently and have a different palate than I do, but we all take the extra time to eat real food instead of grub out of a box or a bag or a bottle. No single food acts as a silver bullet for improving or maintaining brain health despite the superfood halo that some foods get. Remember, it's the combination of foods and nutrients in our meals (the entourage) that likely determines health benefits. To make this as easy and memorable as possible, I've summarized my guide to good eating for the brain using the S.H.A.R.P. acronym.

S: Slash the Sugar and Stick to Your ABCs

You can't argue against the fact that all of us would do well to reduce our sugar intake. It's the easiest way to gravitate toward healthier foods in general and limit the amount of processed junk. The average American consumes 163 grams of refined sugars (652 calories) per day, and of this, roughly 76 grams (302 calories) are from the highly processed form of fructose, derived from high-fructose corn syrup.[16] My guess is that a lot of this sugar intake comes in liquid form—soda, energy drinks, juices, flavored teas, and the like—or we eat it in processed food products. When I eliminated added sugars from my diet after I did a *60 Minutes* piece on how toxic sugar can be to the body, I missed it for a minute, and now have no problem avoiding foods that are typically full of sugar (not to mention other mindless ingredients). It was a win all around. My weight remains stable even during the times when I am not as active, and there is little question how much a high-sugar diet affects the "length of my cognitive day." I can't stay productive as long if I am eating sugar, as the inevitable crash happens.

Sugar intake is related to brain health in a wide variety of ways—too many to detail without probably inducing boredom. Nevertheless, I offer a few reasons why sugar in excess can be so toxic to the brain, and it boils down to our relationship with blood sugar control.

In part 1, I covered how Alzheimer's can now be considered type 3 diabetes, whereby the brain cannot use insulin normally. I also noted that gaining control of blood sugar equates supporting brain health; multiple well-designed studies have found that people with high blood sugar had a faster rate of cognitive decline than those with normal blood sugar—whether or not their blood sugar level qualified as diabetes. High blood sugar can be stealthy in people who are of normal weight, but for those who are obese, it's practically a given. Not only does the excess fat make people insulin resistant, but the fat itself releases hormones and cytokines, proteins that lead to a rise in inflammation, create a slow burning fire in the body and brain, and worsen cognitive deterioration.

When you follow your ABCs (more on this shortly), you will automatically be slashing your sugar consumption and reducing your risk for blood sugar imbalances, insulin resistance, and dementia. I'm not asking you to nix sugar entirely; we all love a little sweetness in our lives. But cutting back on volume and being choosier about our sugar sources is the shift to make. Sugar from a milk chocolate–based candy bar or fruit juice is not the same as sugar from dark chocolate or honeydew melon. When you need to add a touch of sweetness, try a pinch of natural stevia, a drizzle of honey, or a tablespoon of real maple syrup.

And what about artificial sugars? Sorry, but they are not a good replacement. While we like to think we're doing ourselves a favor by replacing refined sugar with substitutes like aspartame, saccharin, or even seminatural products like sucralose, these are not ideal. The human body cannot properly digest these, which is why they have no calories, but they must still pass through the gastrointestinal tract. For a long time, we assumed that artificial sweeteners were, for the most part, inert ingredients that didn't affect our physiology. But in 2014 a landmark paper, which has since been widely referenced, was published in *Nature* proving that artificial sweeteners affect gut bacteria (microbiome) in ways that lead to metabolic dysfunction, such as insulin resistance and diabetes, contributing to the same overweight and obesity epidemic for which they were marketed to provide a solution.[17] These are the same conditions, as you know now, that increase risk for brain decline and serious dysfunction. Try to avoid these sugar substitutes. In general, reducing refined flours and sugars—real and artificial—is a good idea. This means eliminating or severely limiting chips, cookies, pastries, muffins, baked desserts, candy, cereals, and bagels. Watch out for products labeled "diet" or "lite" or "sugar free" because that usually means they are sweetened artificially. Remember that the best foods don't come with nutritional labels or health claims. They are the whole, real foods you find around the perimeter of a grocery store.

Let's get to those ABCs. It's a method to discern the top-quality foods, the A-listers, from the ones we should include (B-list) or limit (C-list). The Global Council on Brain Health, in its 2019 report *Brain Food: The GCBH Recommendations on Nourishing Your Brain,* described the most brain healthy diets from around the world and provided a useful framework for foods to encourage and those to limit. Later in this chapter, I'll give you some ideas for meal making so you can see how these ABCs work in real life; it's similar to the Mediterranean-style diet.

A-LIST FOODS TO CONSUME REGULARLY

Fresh vegetables (in particular, leafy greens such as spinach, chard, kale, arugula, collard greens, mustard greens, romaine lettuce, Swiss chard, turnip greens)

Whole berries (not juice)

Fish and seafood

Healthy fats (e.g., extra virgin olive oil, avocados, whole eggs)

Nuts and seeds

B-LIST FOODS TO INCLUDE

Beans and other legumes

Whole fruits (in addition to berries)

Low sugar, low-fat dairy (e.g., plain yogurt, cottage cheese)

Poultry

Whole grains

C-LIST FOODS TO LIMIT

Fried food

Pastries, sugary foods

Processed foods

Red meat (e.g., beef, lamb, pork, buffalo, duck)

Red meat products (e.g., bacon)

Whole-fat dairy high in saturated fat, such as cheese and butter*

Salt

H: Hydrate Smartly

As we age, our ability to perceive thirst diminishes. This helps explain why dehydration is common in older people, and dehydration is a leading cause for admission to emergency rooms and hospitals for the elderly. A good rule of thumb is that if you feel any thirst, you have already waited too long. (And by the same token, if you feel stuffed, you have already eaten too much.)

One of my mantras is "drink instead of eat." We often mistake hunger for thirst. Even moderate amounts of dehydration can sap your energy and your brain rhythm. Because our brains are not really that good at distinguishing thirst and hunger, if there is food around, we generally tend to eat. As a result, we walk around overstuffed and chronically dehydrated.

The link between hydration status and cognitive ability and mood is well recognized. Dehydration often leads to cognitive problems in older folks, which can be assessed by examining changes in short-term memory, numerical ability, psychomotor function, and sustained attention. Researchers have found that even moderate dehydration is associated with confusion, disorientation, and cognitive deficits.[18] The degree to

---

*There has been a lot of noise around the debate about saturated fat. Which is worse in terms of causing heart disease: saturated fat or sugar? Saturated fat, especially from animal products, is not harmless. If you eat a lot of fatty meats, butter, lard, and cheese, your high intake of saturated fats could raise your risk for all causes of premature death, including dementia. What the research shows, however, is that replacing butter, cheese, and red meat with highly refined carbohydrates (such as white flour products and white rice) does not reduce heart disease risk. I'd rather you enjoy an artisanal cheese plate with whole grain bread or crackers than buffalo wings dipped in blue cheese dressing or chili cheese fries. You get my point.

which thinking skills are affected is dependent on the severity of dehydration, and the extent to which observed cognitive performance and associated neural activity are reversible with rehydration is a topic of ongoing investigation. The lesson here is to stay hydrated, and the best way to do that is with water. You also can have your morning coffee or tea.

Most people get their antioxidant fix in caffeine. Several studies have found an association between drinking coffee and tea and decreased risk of cognitive decline and dementia.[19] We don't know exactly how or why this is the case. We know that the short-term effects of caffeine have been shown to increase alertness and cognitive performance (as well as athletic performance), but the long-term effects are less well understood. Several studies have suggested that those who drink coffee have better cognitive function over time than those who drink less coffee. But it's possible that the caffeine or compounds in coffee and tea may not be the cause of improved outcomes; rather, people who drink tea and coffee are also more likely to have higher education levels or better health, which are tied to improved cognitive performance and lower risk of dementia. The good news is that you're not going to do your brain any harm by drinking coffee or tea unless you're downing copious amounts of caffeinated energy drinks in combination with your coffee (which you shouldn't be doing anyhow). Just be sure your caffeine consumption doesn't interfere with your sleep. For most people, it's ideal to scale back the caffeine intake in the afternoon and be caffeine-free after about 2:00 p.m.

Alcohol does not count as a source of hydration, but it can be part of a healthy diet. We hear competing messages in the news about the benefits (or lack thereof) of alcohol. While there is substantial evidence that moderate alcohol consumption can have protective heart health and cognitive benefits, some studies indicate there are also adverse effects on the brain from alcohol consumption. Alcohol consumed even in moderate quantities has been linked with negative brain health outcomes in some people. And therein lies the difference: *in some people*. For you, a daily glass of wine may help your heart and brain function better over time, but for your friend, the opposite may be true. The trouble with alcohol is that people can slip into abusing it, gain a tolerance to excessive amounts, and establish a bad habit—or worse, addiction. There are both short-term and long-term risks associated with excessive alcohol consumption, including learning and memory problems. Any excessive intake of alcohol will have negative effects on every organ in the body. And as we age, our ability to metabolize alcohol decreases. In 2017, a report published in *JAMA Psychiatry* revealed a jarring trend: alcohol abuse is rising among older adults.[20] The researchers speculate the causes to be everything from increased anxiety in general to more robust late-lifers who think they can continue the drinking habits of their youth.

The debate about the risk-benefit analysis on alcohol—and its related studies—will surely continue, but here's what I suggest: If you don't drink alcohol, don't start drinking in order to protect your brain health. If you drink alcohol, don't overdo it, because it is unclear what the beneficial level of consumption is for brain health. For men, moderation is up to two drinks a day (a drink is 12 ounces of beer, 5 ounces of wine, or 1.5 ounces—one shot glass—of liquor); for women it's one drink. While that is partly because women are physically smaller, more alcohol puts women at higher risk for breast cancer as well. Ideally, choose red wine mostly because it does contain polyphenols, micronutrients that may act as antioxidants that affect blood pressure and are not typically contained in spirits or beer.

A: Add More Omega-3 Fatty Acids from Dietary Sources

We hear a great deal these days about the benefits of omega-3 fatty acids—the brain-nourishing gems from seafood, nuts, and seeds. Unfortunately, the American diet is extremely high in another type of omega—the omega-6 fats, abundant in the corn and vegetable oils used in so much processed, fried, and baked food. The result is that we consume a disproportionate amount of omega-6 fats. According to anthropological research, our hunter-gatherer ancestors consumed omega-6 and omega-3 fats in a ratio of roughly 1:1. Today the average American eats a disproportionate amount of omega-6 fats compared to omega-3s: anywhere from 12:1 to 25:1 omega-6 to omega-3. As you might guess, that's mostly because we eat too much omega-6, while at the same time our intake of healthier, brain-boosting omega-3 fats has dramatically declined from evolutionary norms.

Fatty fish is a wonderful source of omega-3 fatty acids (especially salmon, mackerel, and sardines), and even wild meat like beef, lamb, venison, and buffalo contain this healthy fat. Plant sources of omega-3 fatty acids include flaxseed, plant-derived oils (olive, canola, flaxseed, soybean), nuts, and seeds (chia seeds, pumpkin seeds, and sunflower seeds). Food sources—*not* supplements—are the best way to obtain omega-3 fatty acids. In fact, fish oil supplements have come under scrutiny lately due to mixed results in studies. Although fish oil supplements have been billed as an easy way to protect the heart, lower inflammation, and improve mental health, the evidence is far from definitive and compelling (and yet Americans spend more than $1 billion a year on over-the-counter fish oil).

In January 2019, for example, Harvard researchers reported in the *New England Journal of Medicine* that omega-3, also known as marine n-3, fatty acid supplements did nothing to reduce the likelihood of a heart attack in men fifty and older, as well as women fifty-five and older who did not have any risk factors for cardiovascular disease.[21] Other studies have also shown that taking too much fish oil, which is easy to do with supplements, can have surprisingly negative side effects, such as higher blood sugar levels, increased risk of bleeding due to effects on blood clotting, as well as diarrhea and acid reflux (heartburn).[22] Unless you have a true deficiency, I'd rather you get omega-3s from foods, not supplements. You would be hard-pressed to overdose on fish and walnuts! And remember, nearly all the studies that link omega-3 to brain health have largely been done on food sources, not supplements. That fact alone speaks volumes.

The impact of omega-3 fatty acids on the brain has been extensively studied, and there is a wealth of information on the link between omega-3 fatty acids and healthy brain aging. Studies examining the role of omega-3 fatty acids have widely considered them together as a whole rather than looking at the specific types of omega-3 fatty acids: EPA (eicosapentaenoic acid), ALA (alpha-linoleic acid), and DHA (docosahexaenoic acid). DHA is the most prevalent omega-3 fatty acid in the brain and has been shown to play an important role in the maintenance of neuronal membranes, and fish as well as algae have plenty of it. No surprise, then, that in large-scale surveys, those who typically eat fish or other seafood every week report better brain health than those who never eat fish or seafood.

I think it's safe to say we all would do well to eat more fish. In some regions of the country, fish can be less expensive than meat. Just be sure you know where your fish is coming from. Avoid fish from polluted waters or places where the mercury content in the fish can be too high. Mercury is a heavy metal that can harm the brain and is not easily eliminated from the body. A good resource to check is the Monterey Bay Aquarium's Seafood Watch website (www.seafoodwatch.org). The site can help you choose the cleanest fish (wild or farmed) that are harvested with the least impact on the environment.

MYTH: Supplementing your diet with vitamins, omega-3 fish oil, and vitamin D is a good thing. It will help make up for dietary flaws.

FACT: Supplements do not take the place of real food, and some can be harmful. The supplement industry is woefully unregulated; supplement manufacturers do not have to test their products for effectiveness or safety. While there are some quality supplement makers with a solid and ethical track record, use should be considered on an individual basis under a doctor's recommendation.

Let me talk about supplements more broadly, beyond the fish oils. A basic rule is that when you eat right, you shouldn't need to supplement.

While taking a multivitamin a day might confer a placebo effect (you think it's actually doing you some good or somehow making up for nutritional deficits), they are probably not going to help you prevent any illness or brain decline unless you truly have a nutrient deficiency. Although nutrient deficiencies in the Western world are extremely rare, some neurologists recommend certain supplements based on a patient's individual circumstances and biology. Most of us are eating a fortified diet. Even fresh mushrooms now come "fortified" via irradiation with vitamin D. Researchers, including Pieter Cohen at Harvard, have pointed out that even with a standard American diet, we are probably not likely to be broadly deficient in most vitamins thanks to fortification. The problem is more the quantity of what we are eating, not the deficiency.

When I worked on a film about the supplement industry, I was stunned at just how unregulated it is. Thus far in 2019, the Food and Drug Administration has issued twelve warning letters to companies that were illegally marketing fifty-eight dietary supplements with claims to prevent, treat, or cure Alzheimer's disease or other serious conditions. Supplement makers don't have much of an obligation to prove their product is safe or effective before taking it to market. And as Dr. Dean Sherzai of Loma Linda University and the author of *The Alzheimer's Solution* explains, it is actually harder than you think to take the "good" stuff out of food and put it into pill form.[23] While you may be able to get the active ingredients isolated and even synthesized, real food is made up of a multitude of molecules, and we have only begun to scratch the surface in defining what they all do. Some seemingly inert molecules may help the active ingredients travel through the body, acting as vehicles. Other molecules may help unlock receptors, allowing the molecules to activate their targets. As I mentioned earlier, it is referred to as the *entourage effect* and helps explain why real food is always going to be a better option than a supplement.

Keep in mind that most studies that look at the utility of supplements rely on self-reporting of both usage and symptoms. It leaves a lot of room for interpretation and bias. It is part of the reason we get constantly conflicting studies—one day, it is a great savior and the next, it has no bene-fit. If you contemplate supplementing, do so under the supervision of a doctor. This area needs to be personalized.

MYTH: Taking supplements marketed for boosting brain health, such as ginkgo biloba, coenzyme Q10, and apoaequorin (a protein from jellyfish) is a great way to prevent dementia.

TRUTH: We'd all love to think we can maintain our cognitive powers by popping a few pills a day. These antidementia supple-ments are backed by some clever advertising and are often sold by major retailers, giving them the hue of total legitimacy. But they are not backed by science. No known dietary supplement improves memory or prevents cognitive decline or dementia—no matter what the manufacturers claim in bold promises that you see on the Internet, in newspaper ads, and on TV. These supple-ments are often promoted by testimonials that appeal to people worried about brain health. Don't be fooled. Spend the money you'd waste on supplements on something that will help your brain: a good pair of walking shoes or a new pillow for a good night's sleep.

R: Reduce Portions

You've heard this lesson before: portion control is a potent skill and a potent preventive strategy in any health-related goal. We westerners love our gigantic plates and heaps of food. Look no further than Thanksgiving dinner or the spread on Super Bowl Sunday (as a nation, we eat more food on Super Bowl Sunday than any other day of the year). Occasional overindulgence won't kill you (or your brain), but every day in between, we have to diligently watch our caloric intake. All the experts I spoke with for this book mentioned portion and calorie control. It's a given in any brain wellness conversation.

The easiest ways to gain control of your portions and calories are to prepare meals yourself at home, measure accurately, and don't go back for seconds. You know what you're putting into the meals you cook and have better control over ingredients and portion sizes. The research also speaks: Frequently cooking at home does indeed lead to better diet quality and improved health and weight. One thing we often don't think about, however, is cooking methods and their impact on nutrition. For example, there are benefits of slow, low-temperature cooking such as sautéing as compared to fast, high-temperature cooking such as frying. Frying can generate harmful chemical compounds that may promote inflammation and harm brain health. When possible, turn to boiling, poaching, steaming, or baking. This is yet another reason to cook more at home: You get to decide which method to use. We tend to gravitate toward fried and grilled fare when we're eating out. At home, however, in addition to controlling for cooking methods, you can also avoid those mystery oils, sauces, and added ingredients. If time is an issue and you can splurge a little, take advantage of the increasing volume of home grocery delivery services.

How about fasting? Intermittent fasting has come back into the spotlight in recent years as a method of reducing calorie consumption, another subject matter I encountered a lot when I was researching this book. There are two common approaches to fasting. One is to eat very few calories on certain days, then eat normally the rest of the time. The other involves eating only during certain hours and skipping meals for the rest of each day. I know a lot of fellow doctors who eat just two meals a day and go for extended periods of time without eating. They will fast overnight from dinner until lunch the following day, thereby fasting for twelve to sixteen hours straight. This helps reduce their overall caloric input (unless, of course, they're consuming huge portions when they do eat). Although large, long-term studies on the benefits of fasting are still lacking, there is some evidence in animal models that it can slow the progression of certain age-related diseases and boost memory and mood. It also has been shown to improve insulin sensitivity, a good thing in the name of metabolism and, ultimately, brain health.[24]

Dr. Mark Mattson is a professor of neuroscience in the Johns Hopkins School of Medicine and also serves as chief of the Laboratory of Neurosciences at the National Institute on Aging. He has devoted much of his life to studying the brain and the effects of cutting caloric intake by fasting up to several days a week.[25] In laboratory experiments, Professor Mattson and his colleagues have found that intermittent fasting, which in his definition means limiting caloric intake at least two days a week, can help improve neural connections in the hippocampus while protecting neurons against the accumulation of those dangerous amyloid plaques.[26] According to his theory, fasting challenges the brain, forcing it to react by activating adaptive stress responses that help it cope with disease. From an evolutionary perspective, this makes sense. One thing we know is that when fasting is done correctly, it can increase the production of brain-derived neurotrophic factor (BDNF), a protein I defined earlier that helps protect and strengthen neural connections while also spurring new growth of brain cells. Physical exertion and cognitive tasks can also trigger higher levels of BDNF.

Fasting is not for everyone (it can take some getting used to, sort of like exercise if you've been sedentary), but I'll give you some ideas in chapter 9 should you want to try it and have checked with your doctor. I've tried it several times, and after the first time, it becomes much easier than you might think.

P: Plan Ahead

Put another way, don't get caught starving and resorting to junk food (simple carbohydrates, lack of fiber, and saturated fats). Food is all around us, especially the junky kind. When hunger strikes and we're not prepared, well-ingrained animal instincts will push us in the wrong direction. We will gravitate toward whatever is quick, tasty, and satisfying (hello, cheeseburger, fries, and soft drink).

Once or twice a week, try to plan your main meals in advance and grocery-shop accordingly. Aim to build more fiber into those meals, including whole fruits and vegetables (for fruits, bananas, apples, mangoes, and berries rank high on the fiber content; for vegetables, the darker the color, the higher the fiber content); beans and legumes; whole grains; and seeds, including wild and brown rice. I haven't talked much about fiber, but it's key to brain health because it changes the overall chemistry of a meal. When you lack fiber, the carbohydrates you eat will get absorbed more quickly, thereby raising glucose and insulin levels and potentially increasing inflammation. Fiber intake has long been shown to help prevent depression, hypertension, and dementia through a variety of biological pathways.[27] It also has been tied to successful aging in general. There are two types of dietary fiber: soluble and insoluble. Soluble fiber dissolves in water, which turns it into a type of gel that lowers cholesterol and glucose levels; it's found in oats, peas, beans, apples, carrots, and citrus fruits like oranges. Insoluble fiber can't dissolve—it's the roughage that keeps the other digestive fluids moving through your intestines. This type of fiber is found in nuts, whole grains, wheat bran, and vegetables like green beans. It's the tough matter that's not broken down by the gut and absorbed into the bloodstream (it stays intact as it moves through the digestive system).

There's no easier way to consume more fiber than to plan meals ahead, focusing on adding more fibrous plants to the plate and avoiding the fiberless fare from eating out at generic restaurants or out of a box.

Additional Tips

Organic? Grass Fed?

Contrary to reports in the media, we have no good proof that eating organic foods provides any more nutrition than conventionally grown foods. Most people concerned about organic versus conventional are thinking about how pesticides, herbicides, and trace amounts of hormones and antibiotics can potentially have adverse health effects, even if that has not been adequately proven. When people ask me if it's ideal to eat purely organic, I say that given the current science, in general it's not necessary. But if you're concerned about chemical exposures due to traditional farming practices, try not to buy produce that's on the Dirty Dozen list, published each year by the Environmental Working Group (EWG). This list is based on U.S. Department of Agriculture findings of conventionally grown foods most likely to contain pesticide residues: strawberries, spinach, nectarines, apples, grapes, peaches, cherries, pears, tomatoes, celery, potatoes, and sweet bell peppers. Fruits and vegetables with thicker skins tend to have fewer pesticide residues, because the thick skin or peel protects the inner flesh. Remove the skin or peel, as you would with a banana or avocado, and you're removing much of the residue. The EWG puts out another list of foods too, called the Clean 15: avocado, sweet corn, pineapple, cabbages, onions, sweet peas, papayas, asparagus, mangoes, eggplants, honeydews, kiwis, cantaloupes, cauliflower, and broccoli.

When you want to occasionally splurge on a good steak, grass-fed beef is a better alternative to conventionally farmed cattle. Grass-fed beef, which comes from cattle not fed grains like corn, has a different composition as a result. It has less total fat, more heart- and brain-healthy omega-3 fatty acids, more conjugated linoleic acid (another type of healthy fat), and more antioxidant vitamins, such as vitamin E. Another strategy that works for me is not keeping meat at home; I consume it only when I'm eating out. This helps me stick to a more plant-based diet low in red meat.

Spice It Up

The fare of my Indian heritage is rich in spices. Turmeric in particular is considered one of the seven essential Indian spices, and it's not only a darling in traditional Indian cooking but is gaining star status in research circles as well. Curcumin, the main active ingredient in the spice turmeric, which is the substance that gives Indian curry its bright color, is currently the subject of intense scientific inquiry, especially as it relates to the brain. It has been used in traditional Chinese and Indian medicine for thousands of years. Lab studies have repeatedly shown that curcumin has antioxidant, anti-inflammatory, antifungal, and antibacterial activities, although we don't know exactly how it exerts these effects. Its powers have attracted the interest of research scientists around the world, including epidemiologists searching for clues to explain why the prevalence of dementia is a lot lower in communities where turmeric is a staple in the kitchen.

In 2018, a study conducted at UCLA led by Dr. Gary Small, a top physician and researcher on the aging brain that I introduced earlier, hit the media for its stunning results: People with mild memory problems who took 90 milligrams of curcumin twice daily for eighteen months experienced significant improvements in their memory and attention abilities.[28] They also had a boost in mood. This was a well-designed, double-blind, placebo-controlled study that involved forty adults between fifty and ninety years old. Thirty of the volunteers underwent PET scans to determine the levels of amyloid and tau in their brains at the start of the study and after eighteen months. (Tau proteins, you will recall, are a microscopic component of brain cells that are essential to neuronal survival. But when they undergo chemical changes, they can become damaged, altered, clumped up, and thus harmful.) After the trial, the brain scans showed significantly fewer amyloid and tau signals in regions of the brain that control memory and emotional functions than those who took placebos. As of right now, there aren't any approved medications that can reliably do the same thing. The researchers are embarking on a follow-up study with a larger number of participants.

Turmeric is one of many spices that can lend flavor to dishes. It's one of my favorites, and we use it a lot in my home. In addition to classic spices and herbs, seasonings and condiments are often part of meals. They can be a source of flavor and nutrition, but a note of warning: They can also be combined with sugar, salt, saturated fats, and other ingredients we'd do well to limit. This is especially true when it comes to certain condiments, prepared sauces, and salad dressings. Read the labels.

The Gluten Debate

I'm pretty sure you've heard about gluten or, more precisely, the gluten-free diet. Gluten is the main protein component of wheat, rye, and barley. It's found in many foods, including breads and pastas, cookies, muffins, and breakfast cereals (and it's often the reason these foods have a delectable, chewy texture). You've also likely heard about people avoiding gluten for a variety of reasons, from losing weight to supporting intestinal health. A gluten-free diet is the only proven treatment for celiac disease, an immune-based disease affecting nearly 1 percent of the U.S. population. In people with celiac disease, dietary gluten triggers an immune reaction that results in intestinal damage. These people must avoid gluten, or they can suffer serious health consequences such as abdominal pain and diarrhea and even nonintestinal symptoms like headache, osteoporosis, and fatigue. Anecdotally, many patients with celiac disease report that when they are unintentionally exposed to gluten, they develop recurrent symptoms that often include transient cognitive problems, including word finding and memory difficulties. This phenomenon, often referred to as "brain fog," is not well understood, and the mechanism by which gluten triggers these cognitive symptoms is unknown.

In addition to those with celiac disease, there are also people who describe symptoms including "brain fog" who improve on a gluten-free diet and yet do not have celiac disease. These people are said to have nonceliac gluten sensitivity. Because there is no definitive test to diagnose this condition, it's usually made after a test for celiac disease produces a negative result. Despite popular claims that gluten contributes to cognitive problems in the general population, there is no evidence to suggest that gluten has an effect on mental function in people without celiac disease or nonceliac gluten sensitivity. Given the principle that what is good for the heart is good for the brain, I should note that diets high in gluten have not been linked to heart attack risk. In fact, a low-gluten diet, if it is low in beneficial whole grains, could pose an *increased* risk of coronary heart disease.[29] I also add that people who claim to feel much better when they go gluten-free tend to clean up their diets in ways that do in fact benefit them but have nothing to do with the gluten aspect. They eat wholesome, fresh foods. They engage more in other healthy habits like exercise. And they see outcomes like weight loss and more energy that motivate them to keep doing what they're doing.

You do not have to go gluten free if you don't have celiac. The key is to choose gluten-containing foods carefully. Avoid gluten-containing refined flours found in white bread, crackers, chips, and pastries because they are not doing you much good, and move instead toward more fibrous, whole-grain foods that boost heart and, in turn, brain health.

Feeding Your Brain

- Using smaller plates is an effective method for controlling portion sizes.
- Eat fish (not fried) at least once a week.
- Look at the sodium content in prepared foods you are eating. Baked goods such as bread, canned soups, and frozen foods are typically high in salt, and you might not realize there is a lot of salt in what you are eating.
- Choose frozen vegetables and fruit, which are typically low in salt and high in essential nutrients, when you're creating fresh meals instead of buying frozen, ready-to-eat meals.
- Eat a wide variety of different colored vegetables. The nutrients that give green bell peppers, for example, their color are different from those that give red or orange bell peppers their hue. When you "eat a rainbow" of vegetables, you eat a more diverse array of nutrients, many of them brain-friendly antioxidants. Try to add new vegetables to your diet, and experiment with new ways of cooking and preparing them.

- Use vinegar, lemon, aromatic herbs, and spices to increase flavor in food without increasing salt content.
- Check the labels of spice blends to determine if they contain salt.
- Use mono- and polyunsaturated fats in cooking, such as extra virgin olive oil, canola oil, safflower oil, and sesame oil. For high-heat cooking, try avocado oil.
- Steer clear of partially hydrogenated oils. That's code for trans fats, which are fading out of the food supply but still find their way into lots of processed foods; fried foods like doughnuts; baked goods such as cakes, frozen pizza, and cookies; and margarines and other spreads. Trans fats raise your harmful (LDL) cholesterol levels and lower your good (HDL) cholesterol levels. Eating trans fats increases the risk of developing heart disease, stroke, and type 2 diabetes. All of these diseases can harm the brain and increase your risk of cognitive decline.
- Prepare meals at home. This gives you more control over the salt, sugar, and fat content than if you buy prepared meals or food from restaurants.

My final word of dietary wisdom is *floss*. Dr. Gary Small added this tidbit to our interview, and it's worth sharing here. Flossing—and brushing—your teeth twice daily removes food debris and bacteria buildup that can ultimately lead to gum disease and increased risk of stroke. The connection to the brain? Gum disease entails inflammation. Periodontitis is an infection of the gums, the soft tissue at the base of the teeth, and the supporting bone. As the natural barrier between the tooth and gum erodes, bacteria from the infection have an entry into the bloodstream. Those bacteria can increase plaque buildup in the arteries, perhaps leading to clots. Hence, flossing is now a good-for-brain habit.

CHAPTER 8

Connection for Protection

Let us be grateful to people who make us happy; they are
the charming gardeners who make our souls blossom.

MARCEL PROUST

After Helen's husband died suddenly from heart failure after more than forty years of marriage, her health and cognition declined precipitously in just a few months. Her husband had been her main social companion, and without his presence, she lacked opportunities to interact with others and had very few friends. It had been a long time since Helen had socialized outside the home. She became increasingly isolated and depressed, living alone in a large, cluttered house with nothing much to do other than sit on the couch and watch TV. If her children had not insisted that she move to a retirement community to experience a social network of people and share joint activities, Helen might have continued to deteriorate mentally and met an early death.

The health of one spouse is important to that of the other spouse. The impact of close relationships, particularly marriage, on an individual's health has been investigated from both a physical health and psychological perspective. In the first six months after the loss of a spouse, widows and widowers are at a 41 percent increased risk of mortality. No doubt some of this increased risk is due in part to a loss of companionship. A meaningful relationship with another person brings love, happiness, and comfort to an individual's life. In addition to psychological well-being, however, relationships have been found to be associated with a broad range of other health functions related to the cardiovascular, endocrine, and the immune systems.

There's also plenty of science to back up the fact we need social connection to thrive, especially when it comes to brain health. A look at the data shows that enjoying close ties to friends and family, as well as participating in meaningful social activities, may help keep your mind sharp and your memories strong.[1] And it's not just the number of social connections you have. The type, quality, and purpose of your relationships can affect your brain functions as well. Even your marital status affects your risk. Researchers at Michigan State University found that married people are less likely to experience dementia as they age, and divorcees are about twice as likely as married people to develop dementia (widowed and never-married people have risk profiles in between the married and divorced groups).[2]

It may be that staying social and interacting with others in meaningful ways can provide a buffer against the harmful effects of stress on the brain. I see the anecdotal evidence for this cause and effect every day in my work as a neurosurgeon and out in the field as a journalist. The people I meet who are the liveliest and most joyful and seem to be having a great time despite their advanced age are the ones who maintain high-quality friendships, loving families, and an expansive, dynamic social network. My heart sinks when I meet a patient who has no immediate family and no close friends. There's nothing more heartbreaking than watching someone suffer through a serious medical issue—maybe even facing death itself—alone.

Social isolation and feelings of loneliness are on the rise in our society. It's the paradox of our era: We are hyperconnected through digital media yet increasingly drifting apart from each other and suffering from loneliness because we lack authentic connection. This absence of real connection is epidemic, and medicine is increasingly recognizing it as having dire physical, mental, and emotional consequences, especially among older adults, with about one-third of Americans older than sixty-five years old and half of those over eighty-five now living alone.[3] A new Global Council on Brain Health survey on socialization and the brain health of adults age forty and over found that although most people are at least somewhat socially engaged (with an average of nineteen people in their social networks), a surprising 37 percent said they sometimes lacked companionship, 35 percent found it hard to engage socially, and nearly 30 percent said they felt isolated.[4] Overall, the survey revealed that 20 percent of adults over forty years old were disconnected socially. That's important because adults who said they were happy with their friends and social activities were more likely to report an increase in their memory and thinking skills in the previous five years, while those who were unsatisfied with their social lives reported the opposite—that their cognitive abilities had declined. Dr. Michelle C. Carlson, a professor at the Johns Hopkins Bloomberg School of Public Health in Baltimore, and an issue expert for the Global Council on Brain Health, who participated in the review, calls this "a public health issue," and she's right.

People with fewer social connections have disrupted sleep patterns, altered immune systems, more inflammation, and higher levels of stress hormones. In a 2016 study, isolation was found to increase the risk of heart disease by 29 percent and stroke by 32 percent.[5] Another analysis that pooled data from seventy studies and 3.4 million people found that individuals who were mostly on their own had a 30 percent higher risk of dying in the next seven years and that this effect was largest in middle age (younger than 65).[6] And loneliness in particular has been shown to accelerate cognitive decline in older adults.[7] Those data speak to me. They tell me to pay attention to nurturing my relationships as much as I nurture my health through diet and exercise. Apparently, high-quality socialization is akin to a vital sign.

Neuroimaging studies have been particularly revealing in this new area of brain science. A couple of investigations have been carried out by AARP Foundation Experience Corps, a program that links older adults with kids who are not reading at grade level yet. The program aims to

KEEP SHARP

be mutually beneficial; it helps older adults engage in the community as tutors while children learn the skills they need to do well in school. Remarkably, fMRI imaging showed that the adults who participated in the program improved their cognition over a span of two years and even reversed declines in brain volume in regions vulnerable to dementia (e.g., the hippocampus).[8] Another study, the Synapse Project, also used fMRI in a randomized trial to compare the difference between putting one group of older adults through challenging activities together, such as quilting or digital photography and another group that just socialized.[9] The results? fMRI analysis revealed that those who were engaged in the challenging activities gained improved cognition and brain function that were not seen in the socializing-only group. Finally, the Rush University Memory and Aging Project has shown that those with larger social networks were better protected against the cognitive declines related to Alzheimer's disease than the people with a smaller group of friends.[10] Engaging socially in a larger group, particularly when centered around some sort of challenging activity, seems to be the most protective.

The damaging effects of social isolation start early. Socially isolated children have significantly poorer health twenty years later, even after controlling for other factors. The stories I've uncovered during my own investigative work on loneliness have stopped me cold, in part because I never expected to hear them from the people in front of me—people with no outward hint of a problem, but mostly because the descriptions of their sense of isolation were so upsetting: "It's unceasing, toxic, brutal." "I feel invisible." "It's like living with a hole smack in the center of your chest—a hollow feeling." "My loneliness magnifies every pain in my body." Oprah asked me to talk and write about this for her magazine.[11] At any given time, at least one in five people, or roughly 60 million Americans, suffers from loneliness—and nearly half of Americans always feel alone or left out.[12] They experience both acute bouts of melancholy and a chronic lack of intimacy that leaves them aching for someone special in their life that "gets" them.

The pain of loneliness has really captured my attention. A remarkable study led by Naomi Eisenberger, an associate professor of social psychology at UCLA, found that being excluded triggered activity in some of the same regions of the brain that register physical pain.[13] Feelings of exclusion lead to feelings of loneliness. This makes evolutionary sense because throughout our history, survival has been about social groups and companionship. Staying close to the tribe brought access to shelter, food, water, and protection. Separation from the group meant danger. Loneliness doesn't discriminate; it can affect people who are single and living alone as much as individuals surrounded by people and living in a family unit. And it affects city dwellers as much as people living in rural areas.

"Get Your Moai On!"

In Okinawa, Japan, where an unusual number of elders live past 100 years old (called a Blue Zone—the name of a region with some of the world's oldest people), *moai* ("mo-eye") is one of their longevity traditions. These are social support groups that start in childhood and extend throughout life. The term originated hundreds of years ago as a means of a village's financial support system. Moais were established to gather a village's resources for projects and public works. Someone who needed capital to buy land or take care of an emergency, for example, could turn to the moai for help. Today the concept of a moai encompasses social support networks, a cultural tradition for built-in companionship. People meet to share advice, ask for help, and gossip. Yes, gossip can be a good thing in social exchanges; it's a gateway to a safety net of friends and has been used by humans since our tribal days.

THE SECRET SAUCE TO A LONG, SHARP LIFE

For over eighty years, researchers in the now-famous Harvard Study of Adult Development have been tracking how health is influenced by connections between people. They started recording data in 1938 during the Great Depression, following the health of 268 Harvard sophomores, and what they've found contains lessons for all of us. (Of the original group recruited, only nineteen are still alive; among the original participants were President John F. Kennedy and long-time *Washington Post* editor Ben Bradlee. Women weren't in the original study because Harvard was still all men, but since then, the researchers have expanded the diversity of their recruits and have included the original men's offspring.) The study is currently led by Dr. Robert Waldinger, a psychiatrist at Massachusetts General Hospital and a professor of psychiatry at Harvard Medical School. His TED talk on the subject, "What Makes a Good Life?" has been viewed more than 29 million times.[14]

Dr. Waldinger's findings are attractive because they debunk commonly held myths about health and happiness. The findings are based on a comprehensive review of the participants' lives and biology. They not only answer questionnaires, but their medical records are combed, their blood is drawn, their brains are scanned, and their family members are interviewed. The lesson learned is that health and happiness are not about wealth, fame, or working harder. They are about good relationships. Period. According to Dr. Waldinger, "We've learned three big lessons about relationships. The first is that social connections are really good for us, and that loneliness kills. It turns out that people who are more socially connected to family, to friends, to community, are happier, they're physically healthier, and they live longer than people who are less well connected. And the experience of loneliness turns out to be toxic. People who are more isolated than they want to be from others find that they are less happy, their health declines earlier in midlife, their brain functioning declines sooner, and they live shorter lives than people who are not lonely."[15]

The Harvard Study of Adult Development has also discovered that it's not the number of friends you have, and it's not necessarily whether you're in a committed relationship; rather, it's the quality of your close relationships that matters. In terms of the brain specifically, it turns out that "being in a securely attached relationship to another person in your eighties is protective." As Dr. Waldinger expressed in his TED talk, "One of the key ingredients was that people in relationships where they really feel they can count on the other person in times of need had their memories stay sharper longer. And the people in relationships where they feel they really can't count on the other one, those are the people who experience earlier memory decline. And by the way, those good relationships don't have to be smooth all the time. Some of our octogenarian couples could bicker with each other day in and day out, but as long as they felt that they could really count on the other when the going got tough, those arguments didn't take a toll on their memories."[16]

Dr. Waldinger encourages people to lean into relationships with family, friends, and community. It could be as simple as spending more time with loved ones or reaching out to someone you haven't spoken to in years but has a place in your heart. And you can make new friends however old you are. What happens naturally as we age is that we lose connections due to deaths, challenges with mobility, and geographical separation. Our social networks can shrink from the effects of retirement or an illness. Seeking out new connections can counter those developments.

MYTH: Money and fame will keep you happy throughout life.

TRUTH: Close relationships protect people from life's discontents, help to delay mental and physical decline, and are better predictors of long and happy lives than social class, IQ, financial status, or even genes.

While social media can be isolating, they also present new opportunities for older adults to engage socially when used appropriately. More than 80 percent of Americans, including older adults, use the Internet daily. No doubt this sort of digital engagement should complement rather than replace in-person communication, but email, instant messaging software, social networking sites, online communities, and blogs can help us maintain our relationships with family and friends and expand our social world. Studies of online communities for older persons showed that community members report numerous benefits, including intellectual stimulation, playful experiences, and emotional support.

Such social engagement may be particularly valuable for older individuals living in remote places or unable to get around. To an extent, virtual connections may compensate for lost relationships and offer relief and distraction from stressful circumstances. In addition, thanks to the anonymity, invisibility, and opportunity for reading and responding to communication as schedules permit, digital engagement enables people to more easily communicate with others and get across their feelings, opinions, and skills. We think that this has the effect of instilling more confidence and a sense of control of one's life—all good things for health.

I've seen a lot of disparity in my travels. After the basics, one of the big drivers of disparity is whether or not someone has access to the Internet. Granted, some communities are better off being disconnected from modern technology. I won't be encouraging the closely knit tribes I visited along the Amazon River to start installing Wi-Fi. But there's something to be said for the vast majority of people living in the developed world to stay connected and keep up with acquiring new computer skills. When I meet older folks who learn how to use a computer and tools like email, social media, and search functions, they seem to have a much greater sense of independence and generally appear happier than the people who stay offline. I know that stands in contrast to how a lot of people view technology, but there are plenty of studies to back this up. The Internet affords us many opportunities to learn and connect with others. There's even some evidence demonstrating that digital engagement can have positive effects on cognitive abilities in later life that is on par with in-person communication. An Australian study involving more than five thousand older men found that those who use computers have a lower risk of receiving a diagnosis of dementia by up to eight and a half years,[17] and an experimental study conducted in the United States revealed that older adults performed about 25 percent better on memory tasks after learning to use Facebook.[18]

Here are some tips to staying socially engaged:

- Focus on the relationships and activities that you enjoy the most, such as a team sport, interest groups, or political activities.
- Ask others for help in removing barriers to social interaction— for example, difficulty getting around due to physical limitations or the fact you don't drive anymore.
- Make a point of connecting regularly with relatives, friends, and neighbors. Digital contact is important too.
- Maintain social connections with people of different ages. That means people older and younger than you are.
- Volunteer at a school or community center.
- Look for programs in your community that allow you to pass along skills you have, such as cooking or coaching a team. You can start by seeing what kind of activities are available at your local community rec center or community college.
- Try to have at least one trustworthy and reliable confidante to communicate with routinely (e.g., weekly)—someone you can trust and count on.
- Add a new relationship or activity. Place yourself in everyday situations where you can meet and interact with others (e.g., stores or parks).

- Challenge yourself to try out organized clubs such as travel and book clubs.
- Consider adopting a pet. Caring for a cat, dog, or bird can be a catalyst to social interaction. Taking care of pets can give a sense of purpose and structure to a pet owner's day. Specific benefits to adults interacting with animals range from reduction in depression, anxiety, and social isolation to decreased blood pressure, reduced risk of heart attacks, and increased physical activity. Dogs can be social icebreakers by serving as a conversation trigger between strangers or casual acquaintances. As a result, dog walkers are more likely to experience social contact and conversation with other people than walkers without pets.
- If you are feeling isolated, reach out to professionals who can help you, including religious leaders, telephone hotlines, and therapists.

When I interviewed UCLA's Dr. Gary Small (he's the one who endorses flossing), he suggested the "triple threat": Take a walk with a friend or neighbor and have a conversation about what worries you. The combination of the exercise, in-person interaction, and talking through your anxieties is a wonder drug to the brain. Dan Johnston at BrainSpan added a good point about the foundation of relationships in general: "You have to have a good brain to have good relationships." There's a beautiful circle of success here: Good relationships boost the brain, and a healthy brain boosts relationships.

But as people living with early stage dementia know, your brain doesn't have to be razor sharp to still have good relationships. Too many people worry about the stigma of fading memories or cognitive decline and either begin to withdraw themselves or lose lifetime relationships because their old friends don't know what to say to them. This is the opposite of the virtuous circle described above. It is especially important for those living with dementia and their caregivers to have people reach out to them to keep those relationships going or develop new ones. Remember: You can't catch dementia from others, and sharing a smile and a laugh may be the best medicine there is.

Finally, don't underestimate the power of appropriate touch. Hand holding has been found to decrease levels of the stress hormone cortisol. A friendly touch can also be calming. In other words, the simple act of touching another human is a way of connecting with others to protect ourselves—and them.

CHAPTER 9

Putting It All Together

12 Weeks to *Sharper*

One, remember to look up at the stars and not down at your feet.
Two, never give up work. Work gives you meaning and purpose
and life is empty without it. Three, if you are lucky enough
to find love, remember it is there and don't throw it away.

STEPHEN HAWKING

I had a chance to spend a few days with Stephen Hawking in the late 1990s. I was working at the White House and helped plan a series of evenings for the President and First Lady. When we were thinking about how best to celebrate science, we unanimously agreed on having the famed theoretical physicist as our featured guest. Because of his ALS, Hawking typed (with one finger) his entire talk into his computer and then played it after going onstage. We had even planned the question-and-answer part of the evening ahead of time. I knew the audience would be engaged and entertained by his brilliance in the world of physics, but it was his life lessons that have stayed with me more than twenty years later. Hawking's disease slowly robbed him of his ability to walk, speak, and participate in life the way most of us can, but he had a mind that nobody and nothing could take away from him. It remained sharp until his peaceful death, which happened to be on the 139th anniversary of Einstein's birth.

Since my youth, I have appreciated the idea that we each "own" our brains. And like Hawking, I've never taken mine for granted. When I was a child, my father was mugged and robbed. It was pretty traumatic for everyone in the family, and I didn't realize how much I had internalized the feeling that my family had been violated. It practically felt like *I* had been violated—that the culprit had taken something from me. Once when I was speaking to a teacher about it, he told me (while pointing to his head): "They can take away everything you own, but they can never take away *this.*"

It's true. There will always be bad guys out there trying to steal our possessions and momentarily disrupting our life, but they cannot pilfer our minds. Our minds are uniquely our own, and our perceptions of the world are uniquely our own as a result. As soon as sensory stimuli come in—through smell, sight, sound, touch, taste—they go through hundreds of relay stations, each one changing the stimulus ever so slightly so that the final interpretation of the stimulus is highly individual. This is what makes each of our lives distinctive as well. I plan to continue living a wholly unique life of adventure and discovery for as long as possible. It will create a mind that is like a fingerprint, unlike anyone else's. My hope is the same for you.

I've given you a lot of information in this part of the book, much of it geared toward teaching you strategies that will keep you sharp. Now I offer you a twelve-week outline of a program you can use to put my ideas into daily practice. Never forget that the brain is exceptionally plastic—it can rewire and reshape itself through your experiences and habits, and a lot of this remolding can be achieved in a mere twelve weeks. It's like building any other muscle.

You might be feeling overwhelmed or perhaps panicky at the thought of following this program if it means ditching some of your favorite foods, starting an exercise routine after having been sedentary for a long time, trying to learn to meditate, and getting out of the house more often to socialize. I realize that for some people, breaking an addiction to sugar and breaking a sweat more often can be tough. Change is a challenge, and changing long-established habits takes effort. You're wondering if this is truly doable in the real world if your willpower is weak.

Well, let me say that you can do this. Take the plunge and experience the initial effects. Within a couple of weeks, I predict that you'll have fewer anxious thoughts, better sleep, and improved energy. You'll feel clearer-minded, less moody, and more resilient to your daily stressors. Over time, you'll likely experience weight loss, and specific laboratory tests will show vast improvements in many areas of your biochemistry—from what's going on in your brain to how your metabolism and immune system are functioning.

It's smart to check with your doctor about beginning this new program, especially if you have any health issues such as diabetes. Do not change any of your medications or other doctor-prescribed recommendations. But do think about getting some baseline testing done with your doctor to see where you can reduce your risk from a metabolic standpoint. As I've outlined, things like blood pressure, and levels of cholesterol, blood sugar, and inflammation all factor into risk for cognitive decline. You can fight those numbers and bring them into a healthy range either through lifestyle changes or in combination with certain drugs. This is standard bloodwork done at regular checkups. It can further motivate you. The program outlined below will automatically help you to address these important areas, and I encourage you to re-check your numbers after you've gone through the program. My guess is you'll see improvements.

Take this one day, one change at a time. You do not need to follow this program precisely. All I ask is that you do what you can and aim to establish at least one new habit a week throughout these next twelve weeks.

Over the course of the twelve weeks, you will achieve five important goals:

1. Move more throughout your day and build an exercise routine into your life.

2. Find new ways to stimulate your brain through learning and challenging your mind.

3. Prioritize getting restful, routine sleep at night and incorporate daily destressing practices into your routine.

4. Introduce a new way of nourishing your body.

5. Connect authentically with others and maintain a vibrant social life.

During the first week, you will start five new habits each day based on the five pillars, and then repeat the new series of habits the following week. In the third week, you'll incorporate more habits into your days until you've reached the twelfth week with a whole new rhythm. It may take you a little longer to fully establish and maintain these healthy behaviors for life, but the first twelve weeks will get you started. It's your launchpad. You needn't do anything to prepare; you can start today. Although some planning will be involved, such as scheduling workout times, shopping for my menu ideas, or creating a weekend get-together with friends, you can work these suggestions into your life as you see fit.

I won't be asking you to buy anything to make this program work for you. I'd love for you to invest in yourself in the form of enrolling in a creative writing class, for example, or joining a local yoga studio, but these may not align with your preferences. Tailor this program to your needs and proclivities. If I make a suggestion that you don't like, skip it or replace it with another. I want this program to be flexible, doable, and personalized. Don't second-guess your ability to succeed at this; I've designed this program to be as practical and easy to follow as possible. Most important, it will end up being tailored and highly individualized for you.

WEEKS 1 AND 2: DIVE INTO THE FIVE

You can address five areas in your life in these next two weeks and start building a better brain.

Move More

If you already exercise regularly, keep it up, but try something different this week to surprise your body and use new muscles. If you're a jogger, try swimming or a cycling class. Aim to increase your workouts to a minimum of thirty minutes a day, at least five days a week. Don't forget to engage in strength training two to three days this week, avoiding back-to-back strength training days so you give your muscles time to recover. On the days when you don't want to do anything strenuous, go for a long walk or take a restorative yoga class.

For those who haven't busted a move in a while, time to get moving. If you've been totally sedentary, start with five to ten minutes of burst exercise (thirty seconds of maximal effort and ninety seconds of recovery) and work up to twenty minutes at least three times per week. You can do this any number of ways: walking outside and varying your speed and levels of intensity with hills; using classic gym equipment such as treadmills and StairMasters; or taking online exercise classes at places like Daily Burn or Booya Fitness to perform a routine in the comforts of home (most of these require a fee or monthly subscription, but they offer free trial sessions so you can find the one you like the best). Get out your calendar, and schedule your physical activities.

If you have a day with absolutely no time to devote to a continuous segment of formal exercise, then think about the ways you can get in more minutes of physical activity during that day. There's a general lack of walking, standing, and moving our bodies on a regular basis to counteract all the harm that can result from sitting for the majority of the day. All the research indicates that you can achieve similar health benefits from doing three ten-minute bouts of exercise as you would from doing a single thirty-minute workout. When you are short on time on any given day, break up your routine and think of ways to combine exercise with other tasks; for example, conduct a meeting with a colleague at work while walking outside, or stream your favorite show while you complete a set of yoga poses on the floor. Limit the minutes you spend sitting down. Every time you are about to sit, ask yourself: "Can I stay standing and moving instead?" Walk around while you talk on the phone, take the stairs rather than the elevator, and park a distance away from the front door to your building. Simply make a point to get up every hour for a five-minute stroll or jog in place. The more you move throughout the day, the more your body and brain benefit.

Love to Learn

In chapter 5, I covered the importance of participating in cognitively stimulating activities. How often are you reading books and learning about topics outside your professional interest? Have you wanted to learn a new language? Take a painting or cooking class? Join a writing group to finish that book of yours? Now's the time to make this happen. I don't expect you to sign up right away, but begin to explore the possibilities in your community. Check out the local university's adult education courses, or maybe your local rec center has programs. You can probably do much of this homework online.

Sleep Hygiene

I gave you plenty of tips for establishing good sleep hygiene in chapter 6. If you get fewer than six hours of sleep per night, you can start by increasing that period of time to at least seven hours. This is the bare minimum if you want to have normal, healthy functioning physiology from your brain on down. If you don't know where to start with better sleep habits, focus on the following:

- *Time your last meal smartly.* Leave approximately three hours between dinner and bedtime so your stomach is settled and poised for you to go to sleep. Avoid late-night binges. Stop caffeine after 2:00 p.m.
- *Be ritualistic with your sleep habits.* Go to bed and get up at roughly the same time daily no matter what. In the hour before bedtime, do something calming: take a warm bath or read a book. Keep your bedroom quiet, dark, and electronics free.

In addition, pick one stress-reducing strategy to practice once a day for at least fifteen minutes. It can be deep breathing, meditating, or journal writing. Just fifteen minutes.

Eat Sanjay Style

I try to eat only when the sun is shining. Some have called this chrono eating—"chrono" meaning relating to the body's sense of time and its circadian rhythm throughout the twenty-four-hour solar day. I believe *when* you eat is also important, not just *what* you eat. I eat breakfast like a king, lunch like a prince, and dinner like a peasant. Frontloading calories seems to help me, and studies have shown we tend to eat less overall as long as we are consistent. I rarely snack, which for most people is just a form of recreational or comfort eating.

When I lived with the indigenous Tsimané tribe in the Amazon rain forest for several days in the summer of 2017, it was one of the wildest experiences of my life. From La Paz, Bolivia, we flew into Rurranbaque, a little town on the edge of the Amazon. From there we took a four-by-four as far as we could go into the rain forest. Then we got into dugout canoes and spent hours on the rivers and tributaries of the Amazon until we found the tribe. I went on the journey because I had heard the Tsimané tribe, as they are called, has almost no evidence of heart disease, diabetes, or dementia. It is an extraordinary thing, considering that in the United States, we spend a billion dollars a day on heart disease, and it remains the biggest killer of men and women alike. In the middle of the Amazon, they didn't even have a health care system, and they seemed to have found something that had eluded us in one of the wealthiest countries in the world. I was determined to learn about their secrets to health. I went spearfishing with one tribe member who thought he was eighty-four years old, but he really didn't know for sure. His shirt was off, and he was balancing himself on the canoe, just looking at the water, spearing fish. His eyesight was perfect, as was his hearing. The entire indigenous tribe was very much like him. What I found is that they typically eat 70 percent carbs (unrefined and unprocessed), 15 percent fat, and 15 percent protein, a percentage I strive for as well.

Members of the Tsimané tribe typically walk (not run) about seventeen thousand steps per day, they rarely sit, and they get nine hours of sleep every night—waking up to the rooster's crow. To be clear, their life expectancy isn't higher because they tend to die of trauma: accidents, snake bites, childbirth, etc. But until the day they die, they are typically very healthy.

When you start the Sharp program, avoid eating out during the first two weeks so you can focus on getting the dietary protocol down. This will prepare you for the day you do venture outside your home for a meal and have to make good decisions. The first two weeks will also dramatically curb your cravings, so there's less temptation when you're looking at a menu filled with brain-busting foods. When you're time-starved and don't have access to a kitchen, often the case during lunch at work, pack food. Remember S.H.A.R.P. (see chapter 7 for details):

S Slash the sugar.
H Hydrate smartly.
A Add more omega-3s from natural sources like wild cold-water fish.
R Reduce portions.
P Plan meals ahead.

Below are some ideas for meal making:

Building a Better Breakfast

Instead of pastries, doughnuts, bagels, or cereals, try one of the following:

- Egg frittata with lots of colorful veggies and a side of whole-grain toast topped with almond butter
- Steel-cut oatmeal with cinnamon, blueberries, crushed raw walnuts, and a drizzle of honey
- Greek-style yogurt (plain, 2 percent) topped with flaxseed, fresh berries, and a tablespoon of real maple syrup (not the kind made with high-fructose corn syrup)
- Whole-grain waffles or pancakes with blueberries and crushed walnuts, topped with a tablespoon of real maple syrup

Skip the juices, smoothies, and frappuccinos, and choose a tall glass of water, black coffee, or tea instead. I don't drink a lot of juices and smoothies in general, despite their popularity. Given that digestion begins in the mouth, juices or smoothies—even super-healthy ones—don't get absorbed as well because they pass through the stomach and first part of the small intestine before digestion really begins. As a result, you are not getting the "good stuff" out of food as easily. Remember that we drink to hydrate more than to obtain calories or nutrients. I stick to real food for these things.

Over the last few years, I have been "drinking" a chewable juice called Chuice (there are a few other brands on the market). With nuts and whole plants in the drink, you are forced to chew your juice, release salivary amylase, and begin the digestion process. Because you are chewing, your stomach and digestive tract are prepared for this bolus of food, and absorption is much more efficient and complete. So if juices and smoothies are your thing—and they can be very helpful when you have on-the-go mornings—look for one of these chewable varieties. Make sure it's low in sugar.

Smarter Lunching

Instead of hitting a fast-food drive-through or buying a highly processed lunch, try these:

- A leafy green salad with lots of colors and a serving of healthy protein such as chicken, salmon, or tofu, topped with seeds, nuts, a drizzle of extra virgin olive oil, and balsamic vinegar
- A turkey or grilled chicken sandwich on whole-grain bread or sourdough with a side of leafy greens

Swap out your daily fix of soda or sugary energy drinks and go for water, unsweetened tea, or try kombucha tea. For a touch of sweetness after lunch, have a serving of fruit or two squares of dark chocolate.

My Kind of Dinner

Again, avoid the fast-food options and put some effort into lively conversation around the dinner table with friends or family. Try these:

- Turkey chili with a veggie-rich side salad
- Grilled fish or chicken with spices of your choice (I'd do the turmeric—you know that!) and sides of roasted veggies and wild rice
- A simple pasta dish with homemade pesto and side salad

Stick with water to drink, and if you choose to, you can add a glass of wine, preferably red. See if you can go without dessert.

Bonus: If you've gotten the green light from your doctor to try intermittent fasting, you can do a mild version once or twice a week by making sure you stop eating after 7:00 or 8:00 p.m. and then going until 9:00

or 10:00 the next morning, respectively, before your next meal. That's a twelve-hour fast, much of which you'll spend sleeping. A more rigorous version is for sixteen hours, and that can be done by skipping breakfast entirely. But again, be sure you can do this given any personal health issues. If you have diabetes, be sure to get some guidance from your doctor.

Connect with People

Chapter 8 listed several ways you can elevate your social life. Kudos to you if you already consider yourself a socially active individual. Keep it up. For those who feel more isolated, make it a goal to call someone you haven't spoken with in a while and invite a friend over for dinner.

WEEKS 3 AND 4

Add more to your new routine by choosing at least two of the following options:

- ☐ Go for a twenty-minute power walk after lunch most days of the week.
- ☐ Invite a neighbor over for dinner.
- ☐ Have at least two of your meals each week feature cold-water fish like salmon or trout.
- ☐ Download a meditation app if you haven't done so already, and start to use it daily.
- ☐ If you're still drinking soft drinks, diet or regular, try to eliminate these from your life and switch to just water. You can drink carbonated, flavored water as long as it does not contain any sugar or artificial sweeteners. In the morning, coffee and tea are fine.

WEEKS 5 AND 6

Add more to your new routine by choosing at least three of the following options:

- [ ] If you haven't tried to keep a gratitude journal yet, start one. Each morning, spend five minutes making a list of at least five people or situations you are grateful for. If weather permits, do this outside in the fresh air and morning sunlight. If you repeat items on your previous list, that's okay. Think of things that happened the day before that can be added to the list. They can be as minor as being grateful that you felt pretty good and reached your goals for the day.
- [ ] Add fifteen more minutes to your exercise routine.
- [ ] Try a yoga or Pilates class or go on a hike with a friend.
- [ ] Avoid all processed foods.
- [ ] Add a relaxing activity to your bedtime routine such as taking a warm Epsom salt bath or engaging in some mindfulness meditation during which you simply sit in a comfortable, quiet place and take notice of your thoughts and feelings. That's it! No judging, no problem solving, no list making—just a few quiet moments of stillness and focusing on your breath.

WEEKS 7 AND 8

Add more to your new routine by checking off all five of the following ideas:

- [ ] Look for opportunities to volunteer in your community or at your children's or grandchildren's schools. Find the time. It will be worth it.

☐ Explore your local farmers' market and buy fresh foods.

☐ Schedule a checkup with your doctor if you haven't had one within the year. Be sure to address your current medications, if you take any, and speak candidly about your risk factors for cognitive decline.

☐ Write a handwritten letter to a younger loved one in the family, describing something you've learned in your life that you can pass down as an important lesson.

☐ Read a book in a genre or subject area that interests you but that you're not used to reading. If you typically read mystery thrillers but loved the play *Hamilton*, try reading the namesake biography by Ron Chernow.

WEEKS 9 AND 10

At this juncture, I encourage you to ask yourself the following questions and adjust accordingly based on your answers:

☐ Am I getting at least thirty minutes of exercise at least five days a week and including strength or resistance training at least two days a week?

☐ Am I learning something new that challenges my mind and demands developing different skills?

☐ Am I getting more restful sleep on a regular basis and managing stress better?

☐ Am I following the S.H.A.R.P. dietary protocol?

☐ Am I connecting with friends and family members regularly?

If you cannot answer these questions in the affirmative, go back and read the chapter that covers the area in particular and see if you can make the necessary tweaks to your lifestyle. If you're still not getting results, it may be time to seek professional help. For example, if your sleep is still troubling you, ask your doctor about a sleep study and be sure any medications you take are not interfering. If chronic stress is an issue or you think you might meet the definition for depression, seek a qualified psychiatrist or therapist, or both.

Your environment is more influential to your habit making and habit keeping than anything else, genes included, so pay particular attention to it. In 2019, two promising clinical drugs trials that were in phase 3 for treating Alzheimer's disease were abruptly halted when they were not shown to benefit patients any better than placebo. The drug was supposed to remove damaging beta-amyloid plaques. It was a blow to hopes, yet it once again underscored the complexity of the disease and the fact that we may not be able to rely on a miracle cure through drugs to save us. What will save many of our brains from disease, however, is a focus on prevention and the various things we can control within our environment to foster superior brain health. Take a look around you and where you spend the most time. Is it conducive to living a healthy life?

WEEK 11

During this week, think about how you'd want your family members to deal with a diagnosis of dementia, including Alzheimer's disease. This is a sensitive subject and not something any of us wants to consider. But it's important to have these conversations in advance so we're prepared. As Maria Shriver reminded me, a disease like Alzheimer's is an emotional, financial, and physical journey. Talk to your kids. Write down your wishes, and be as explicit as possible for the what-ifs. I'll be giving you more ideas in part 3 for addressing this area and knowing what your options could be.

WEEK 12

Congratulations. You are in the final week. Make a list of all the things you've done differently these past several weeks and ask yourself: *What worked? What didn't work? Where can I improve?* Then use this week to plan ahead. Take a brisk walk with a friend and discuss what is bothering you.

Create nonnegotiables that you will commit to regularly, such as engaging in physical exercise every day, being in bed the same time every night, and eating according to the S.H.A.R.P. plan. Consider apps that help you track how many steps you take a day and how well you sleep. These tools are not for everyone, but you might find a few programs that ultimately help you maintain a brain-healthy lifestyle. Remember to be flexible but consistent. When you momentarily stray from the program, don't be judgmental and simply get back on. Find goals that can be huge motivators, and write them down. It can be anything from walking or running your town's 10K to planning an ecotour trip with your family. People who decide to focus on their health often do so for specific reasons, such as, "I want to be more productive and have more energy," "I want to live longer without illness," and "I don't want to die in the way my mother did." Keep the big picture in mind at all times. This will help you not only maintain a healthy lifestyle but also get back on track if you occasionally cheat. It's cliché, but it's true: Progress is better than perfection.

PART
3

THE DIAGNOSIS

WHAT TO DO, HOW TO THRIVE

An Alzheimer's diagnosis stirs more fear than any other major life-threatening disease, including cancer and stroke, according to a survey by the Marist Institute for Public Opinion. Each of us at some point will know someone who is living with a form of dementia, be it a family member, friend, or oneself, and the diagnosis will likely be the most devastating that person has ever received. At the time someone hears the news, the awful statistics around Alzheimer's really start to set in. There is no cure, and no new drugs to treat symptoms of dementia have been approved for fifteen years, as 99.6 percent of drug trials terminate in failure and well over four hundred dead ends have cost billions of dollars. (The FDA continues to review experimental drugs, and one could have been approved by the time you read this.)

We've known about Alzheimer's for more than a century and cannot treat it easily, let alone cure it. It's a difficult, complex disease that remains a killer. Dementia also takes a devastating emotional, financial, and physical toll on the families of those who are diagnosed with it. In 2016, nearly 16 million family members and friends provided more than 18 billion hours of unpaid caregiving assistance to those with Alzheimer's and other types of dementia.

That's all bad news, no question, but while writing this book, many people reminded me of the signs of hope that are starting to emerge. Remember, every form of cancer was incurable forty years ago, but now people are surviving. In 1981, HIV came on to the scene and even that's now survivable—some would say close to curable. Researchers strongly believe that we will see not only new treatments for dementia in the near future but also novel diagnostic approaches to detect problems early and intervene much sooner for better outcomes. They believe there may be some sea changes coming that will improve both the length and quality of life for those living with dementia. Dementia doesn't have to be a conversation ender; the old notion of "diagnose and adios" needs to be reframed. Life doesn't end for people with dementia. Much to the contrary, many people can find renewed purpose and a zest for life after the diagnosis, though most do have to go through a grieving period as they accept their diagnosis and plan for their future. That future can feel like the great unknown that involves a lot of uncertainty. Everyone's journey is different, but everyone can personalize it to match their unique needs and resources.

In this final part of the book, I turn to the challenges with diagnosing and treating brain disease, particularly forms of dementia. I also offer solutions to leveraging what we do know to best manage these difficult diagnoses and continue to live a fulfilling life. Dementia does not have to be a death sentence or feel like one for either the patient or the caregivers. My hope is to leave you hopeful. Only ten years from now, the first millennials will be turning forty-nine, Gen-Xers will begin turning sixty-five, and the first boomers will be turning eighty-four—an age at which dementia is most prevalent. The time has come to end this disease.

CHAPTER 10

Diagnosing and Treating an Ailing Brain

This increase in the life span and in the number of
our senior citizens presents this nation with increased
opportunities: the opportunity to draw upon their skill
and sagacity—and the opportunity to provide the respect
and recognition they have earned. It is not enough for a
great nation merely to have added new years to life—our
objective must also be to add new life to those years.

JOHN F. KENNEDY

When I started in the world of journalism, I thought I'd be reporting on health policy and the direction of our health care systems. It was the sort of work that I had done at the White House and formed the basis of much of my writing earlier in my career. As much as I have planned my life, though, my pivotal moments have happened suddenly and completely unexpectedly. I started at CNN in August 2001, and three weeks later, the tragic attacks of 9/11 happened. Immediately, I was the only doctor working at an international news network during the unfolding crisis. Shortly after that, I was covering the conflict in Afghanistan, the anthrax attacks, and the war in Iraq. It was a case of professional and personal whiplash.

Having come from a tiny town in rural Michigan, and having had no exposure to war zones or the military, it was a challenging experience to be completely immersed in a foreign world where the stakes were so high and personal safety was a real concern. I was instantly struck by the first responders, nurses, and doctors who so often rushed in to save other people's lives while putting themselves in the line of fire. To this day, I will never forget the first time I saw that complete and genuine selflessness. The people they were saving were typically total strangers and sometimes captured enemies, and yet, they would say to themselves, *Today I am willing to risk my life to save someone I don't even know*. It remains the most human story I covered. I made a commitment then to always report on the stories of the first responders, and over the past two decades, it is the reason I have covered just about every war, natural disaster, and outbreak in the world. Even in the midst of total devastation and darkness, I've wanted to tell the stories of the bright lights that remind us of our humanity.

Writing this book about brain health has been no different from writing about my experiences out on the battlefield or in an area devastated by a disaster. When it comes to dementia, we are at war. Some people bristle at metaphors that conjure up battle. But I've witnessed the disease cause as much devastation and darkness in families as any other type of calamity. There are numerous casualties when it comes to any neurodegenerative disease. Not only does the individual patient suffer; so does everyone else around him or her—from family members and friends to additional caregivers brought in to help (many of whom volunteer). It's emotionally and physically draining. And then there are the costs in time and money. Adding to the toll is the sheer frustration from a general lack of progress in research circles to arrive at a cure. Victims languish in the limbo of a long-drawn-out disease that can go on for years or even decades with no hope for a cure. Conversations always teeter awkwardly between hope and honesty. But as I describe later in the book, the approach to treatment for dementia is beginning to change. The conversation no longer needs to be solely one of desperation. Instead, we can focus on improvements in care and reshape the experience—particularly with early diagnosis and interventions—showing those with dementia and their caregivers that it is possible to live well with the disease until the elusive cure is found.

I recently had a chance to sit down with Bill Gates and talk to him about research for Alzheimer's disease. He wanted to tell me about his personal financial commitment to finding a treatment or cure. It turns out that losing his memory is one of his biggest fears, which again is the case for most people. One of the areas we discussed the most was the trajectory of brain research and how we could better shape it. With Alzheimer's in particular, a great deal of energy has been spent on trying to find the cure. This is understandable, but it also means resources have been siphoned from simpler goals, such as early detection and coping strategies, which are also important. Remember that the accumulation of amyloid begins in the brain decades before symptoms appear. Unfortunately, that also means the disease is fairly developed and much harder to treat when the patient finally develops clinical symptoms. It does, however, pose opportunities to prevent the disease from becoming symptomatic even if it is not cured. In essence, it would be "asymptomatic Alzheimer's." I am very excited about this possibility. In the world of neurosurgery, we always remind ourselves that our objective is not to make a patient's brain scan look better; we are trying to make the patients themselves better. The point is that even if someone has amyloid plaques in their brain but doesn't have memory loss or other symptoms, that would be a very desirable outcome. In fact, we know there are many people who have amyloid and tau in their brains who never go on to develop the symptoms of dementia. Science is just beginning to explore why that is, but the evidence behind healthy lifestyle choices delaying onset or reducing the severity of symptoms supports a strategy to reduce your risk for dementia right now. So as a starting point, I would make sure the trajectory of brain research is centered on patients first, even if it doesn't mean a home-run new therapy. Yes, patients want an effective treatment or even a cure, but incremental achievements in brain research should be pursued and celebrated as well.

Writing this book made me realize just how much the field of scientific research can suffer from groupthink. Once an established scientist has put forth a theory and received funding to prove that theory, many other labs start to follow behind. The problem is that most of those trials then focus on the same mechanism, which in the case of Alzheimer's research historically has been addressing amyloid in the brain. (We have also seen this with HIV trials. At one point, nearly a dozen of the largest and most expensive clinical trials in the world were all basically trying to prove the same thing. They turned out to be wrong.) With AARP and Bill Gates investing in the Dementia Discovery Fund, which is the world's largest venture fund focused entirely on discovering and developing revolutionary therapies for dementia, there will be more renegade approaches to Alzheimer's that will get a fair chance, such as the idea that the glial cells may be activating the immune system, or that the energy life span of the brain cells may be contributing to the disease, or something else entirely. AARP contributed $60 million and inspired United Health and Quest Diagnostics to pitch in. Perhaps most importantly, there needs to be a research platform that allows sharing of data—otherwise everyone may be pursuing the same or wrong theories—and a safety net that allows researchers to take more chances. People are working right now on this particular issue, including Gates, the National Institute on Aging, AARP, and others.

As a neuroscientist and journalist, I often get to spend time with the people behind all the numbers. It's important to do that and truly understand what life is like for someone with Alzheimer's disease. Sometimes the experience is surprising, but it always informs opinions and thoughts about how best to approach this disease. Sandy Halperin in particular did that for me.

BRINGING HOPE

"All we really are is our thoughts and our brain," Sandy told me. It was the spring of 2013, and he was living an independent life as a resident of a retirement community in Florida with his wife, Gail. That was surprising because he had been diagnosed with early-onset Alzheimer's disease in 2010 when he was sixty. What he couldn't possibly have known was that he was likely just thirty-five years old when his brain began its slow descent into the illness. This is important because when Sandy was diagnosed after he started losing words and forgetting his intentions, the disease was already advanced. Sandy admits that his symptoms had started quietly a couple of years before his formal diagnosis, but he was hesitant to acknowledge them and his family didn't notice the signs.

Ignoring symptoms and waiting to seek medical help is not uncommon. Data from the Centers for Disease Control suggest that nearly 13 percent of Americans reported experiencing worsening confusion or memory loss after age sixty, but most—fully 81 percent—had not consulted with a health care provider about their cognitive issues.[1] For most people, the memory lapses are just that, but it is still worth raising the issue with your doctor. If it is the beginning of Alzheimer's, the clock has already been ticking for years. You don't want to lose more precious time that can be used to intervene with strategies, sometimes in combination with drugs, to slow the ailment's progression and ease some symptoms.

I followed Sandy for several years as his malady unfolded. He was brave to open his home and heart to me and my team so we could witness firsthand what it's like to be diagnosed with such a crushing illness and not know what the future holds.

"There is no pain," Sandy told me in 2016. I had asked about this because of recent papers showing that inflammation in the brain is a primary enemy at the time Alzheimer's disease starts to show itself. Sandy had to search hard for the right words. He said it felt like cotton stuffed deep into the front of his head. He began to eloquently describe this feeling with the precision of a Harvard assistant dental professor, which he once was. But then he stopped because he completely forgot what we were discussing. He looked at me blankly. "Front of your head," I gently prompted him. "Right," he remembered. And for just a few minutes, Sandy would be lucid again.

Sandy also opened up his life and his brain to science. He wants to be a part of the transformative advances taking place in the understanding and treatment of Alzheimer's, even if he is not around to benefit from them. Unwilling to be relegated to the sidelines with his diagnosis or fulfill the image of a person wasting away in a nursing home, Sandy became an advocate for increased funding and decreased shame. He also rallied for patients to remain as active and social as possible, just like him. He maintained a LinkedIn network of Alzheimer's patients, advocates, and physicians for as long as he could until his health finally compelled him to pass the baton. Sandy's story may not have a good ending, but he will have a great legacy.

It was while Sandy was working at the Florida Department of Health that he noticed a memory problem worse than losing his keys and forgetting names. His job was to review dental cases for attorneys and consider the merits of patient complaints before the department. Then he would give a written or verbal report. The job required attention to details. Then one day his memory of a case that should have been fresh in his mind simply vanished. As this forgetting started to occur more and more frequently, Sandy struggled to cope. When an attorney entered his office to discuss a case, Sandy would make an excuse to meet a few minutes later so he could try to refresh his memory. That desperate scramble wasn't going to last. No longer could he hide his symptoms.[2]

At this writing, Sandy is in the more severe stages of the disease, managing as best he can with symptoms and chronic pain from other health challenges. He's supported by a family that includes his wife of more than forty years, his two grown daughters, and his granddaughters (60 percent of family caregivers are women). Among the powerful lessons he wants to leave behind is this: "We're all terminal. . . . I may pass faster, but I have to live my life for now. So I want people to know there's still a good life for anyone with a diagnosis with dementia. That's what they have to know: There's a quality of life they can still have."

That stuck with me. Too many people give up on life when they receive a diagnosis. But you'd be surprised by how much hope and optimism can play into health and any prognosis. In all my years of doctoring and reporting, I've noted that the people who live better—and longer—are the ones who hold on to hope. They keep their chins up and often throw themselves into the service of others. That is what Sandy Halperin did.

A POUND OF PREVENTION

The key to treating dementia is prevention, and it just so happens that the same things you can do to reduce your risk for the disease are what you can do to improve your quality of life as you live with the disease. It's one of Dr. Richard Isaacson's most salient points: Alzheimer's disease typically begins in the brain up to twenty to thirty years before symptoms develop. I've already called out this fact a couple of times because it's so important (and it forces us to think of our children). This presents an opportunity to intervene and delay or even prevent Alzheimer's disease altogether. Remember this because the gap of time between brain changes occurring and symptoms surfacing was mentioned by every expert I spoke to while doing my research. It's called the preclinical time, and it is where Dr. Isaacson and many others have started to train much of their attention.

As I mentioned earlier, at the annual Alzheimer's Association International Conference in 2019, Dr. Isaacson presented a paper that is one of the first to show how his revolutionary lifestyle intervention studies lead to improvements in as little as eighteen months. His programs are designed for each individual based on that person's medical screenings and evaluations, but they all involve similar strategies that target different areas in life that are modifiable. These include attention to diet, exercise, sleep, supplements and drugs when necessary, intellectual stimulation, and stress reduction—all the strategies I outlined in part 2. The brains of people who showed no signs of brain disease at the start of his program can act up to three years younger by some measures after his program. Most important, he has shown measurable improvements among those who have memory loss and have already been diagnosed with Alzheimer's disease. He believes he's helping them turn back the clock. And those who have signs of disease in the brain but no symptoms yet could be delaying their illness by years. If you can't entirely prevent the illness from developing, then at least you can delay it for as long as possible. As mentioned in chapter 7, the average number of personalized recommendations given to the 176 people in his study, who ranged in age from twenty-five to eighty-six, was twenty-one. Some of the recommendations were decidedly simple: eat certain types of fish, add more berries to your diet, and get into an exercise routine. These are nature's "drugs" to beat disease. And the fact people who were already showing signs of mild cognitive impairment could improve by following just 60 percent of the protocol speaks volumes.

All of these participants have a history of Alzheimer's disease in the family though no or minimal (mild) cognitive complaints at the start of the study. He calls his approach the ABCs of Alzheimer's prevention management: A is for the anthropometrics like body fat percentage and muscle mass; B is for blood biomarkers such as cholesterol and inflammation levels, blood sugar, as well as genetic tests; and C is for cognitive performance to test for memory, processing speed, attention, and language. From there, he designs individual protocols and reassesses people's ABCs every six months and makes adjustments accordingly.

For Dr. Isaacson, as it is for me in my work, patient outcomes are the most important evidence of science. He says: "As a clinician with a family history, what I'm doing that is so disruptive is taking the time to get to the heart of the matter here and crafting a plan. People who correct their underlying biology first will be much more likely to respond to typical therapies. You don't have to be a neurologist to do this. Any doctor should be able to do this." And he's right. None of us should have to visit a high-profile clinic to receive basic training in preventing brain decline or even applying fundamental lifestyle habits to delay the onset of a disease or, at the very least, improve symptoms that are already evident. Many of the same interventional strategies that Isaacson endorses and "prescribes" in his clinic are the same ones you can benefit from by reading this book.

Dean Ornish is also trailblazing this path. You'll recall I mentioned his randomized controlled trials currently in motion with colleagues at UC San Francisco to see how lifestyle interventions can change the course of Alzheimer's disease. He goes so far as to use the word *reverse* when talking about the possibilities of nipping the illness in the bud during its early stages. His program is not too different from the twelve-week Keep Sharp plan and includes: a whole foods, low-fat, low-sugar, plant-based diet; moderate exercise; stress management techniques such as meditation; and psychosocial support. He has also recruited the help of other scientists to create a full picture during the study. Renowned geneticist David Sinclair at Harvard is measuring changes in gene expression; Dr. Rob Knight's lab at UC San Diego is tracking changes in the microbiome; Dr. Elizabeth Blackburn's lab at UC San Francisco is documenting changes in telomere length, which are chromosomal sections related to aging; and Dr. Steve Horvath at UCLA is measuring changes in the DNA clock. All of these data will help us further understand a disease as complex as Alzheimer's and point us in new directions for therapies and preventive tactics.

Should I Get Tested for the "Alzheimer's Genes"?

Various genes are known to raise risk for Alzheimer's disease. Although a quarter of Alzheimer's patients have a strong family history of the disease, only 1 percent or less directly inherit a gene mutation that causes early-onset Alzheimer's, also known as familial Alzheimer's disease. These people can show signs of the disease in their thirties; many of them choose to enter clinical trials to help scientists better understand the disease in general. Another set of genes can raise risk for the more common late-onset type of Alzheimer's, but they are not deterministic—that is, just because you carry them does not translate to a diagnosis in your life. One of the most common ones is the APOE gene, which has three types or alleles: APOE2, E3, and E4. Everyone has two copies of the gene, and the combination you're born with determines your APOE genotype—E2/E2, E2/E3, E2/E4, E3/E3, E3/E4, or E4/E4. The E2 allele is the rarest form of APOE and carrying even one copy appears to reduce the risk of developing Alzheimer's by up to 40 percent.[3] APOE3 is the most common allele and doesn't seem to influence risk. The APOE4 allele, present in approximately 10 to 15 percent of people, however, increases the risk for Alzheimer's and lowers the age of onset. Having one copy of E4 (E3/E4) can increase your risk by two to three times, while two copies (E4/E4) can increase the risk by twelve times. While there is a blood test for APOE4, the strongest risk gene for Alzheimer's, this test is mainly used in clinical trials to identify people at higher risk of developing the disease.

Carrying this gene mutation indicates only a greater risk, however; it does not indicate whether a person will ever develop Alzheimer's or already has Alzheimer's. When I ask experts whether it's a good idea to know one's APOE genotype, I receive mixed responses. Some say it's best to know what risks you bear so you can do what you can to prevent the disease. But others suggest that such knowledge can be difficult to shoulder if it's bad news and you don't have proper genetic counseling to support you. Genetic counseling means working with someone knowledgeable in interpreting genomic test results and the probabilities you face with various health risks. (These are health care professionals; your doctor can recommend someone.) Personally, I'd want to know my genetic status, but would recommend that anyone who gets tested do so under the guidance of a physician and counselor. And I'll say it again: Your lifestyle habits will influence the fate of your brain more so than your genetics.

THE THREE STAGES OF ALZHEIMER'S DISEASE

Sandy Halperin's experience highlights an important message: If you're destined to develop a disease like Alzheimer's, time is of the essence. As with a cancer diagnosis, you don't want to wait until the ailment has reached its later stages and interventions to delay the progression are practically useless. The disease typically evolves slowly in three general stages: mild (early stage), moderate (middle stage), and severe (late stage). Sometimes the stages are broken down into seven phases, from 1 (no impairment) to 7 (very severe decline), but here I'm going to cover how the Alzheimer's Association categorizes the progression of the disease.[4] Everyone with Alzheimer's experiences these stages differently. How fast and how severe one person shows symptoms and moves through the stages may not reflect someone else's experience, so there's no way to tell how a given disease will advance. That unknown alone can be scary. On average, a person with Alzheimer's lives four to eight years after diagnosis but can live as long as twenty years, depending on other factors. Unfortunately, many people don't get diagnosed until later stages, if at all. This can be especially true for those who are single or otherwise lack partners to notice cognitive and behavioral changes and lapses in memory. And remember: Alzheimer's disease is not the only form of dementia. As I've outlined, the symptoms of other forms of dementia can be different and people can have mixed dementias. These other diseases have similar stages.

Let's review the stages as defined by the Alzheimer's Association.

Early Stage: Mild Alzheimer's Disease

In the early stage, a person may function independently. He or she may still drive, work, and socialize normally. But the person may begin to notice unusual memory lapses, such as forgetting familiar words or the location of everyday objects. Friends, coworkers, family, or others begin to notice difficulties too. This state is also referred to as mild cognitive impairment, particularly when the cause of the dementia is not known. Doctors may be able to detect problems in memory or concentration by asking certain questions. Common difficulties include:

- Struggles with finding the right word
- Problems with remembering names when introduced to new people
- Difficulties performing tasks in social or work settings
- Forgetting material that one has just read
- Losing or misplacing a valuable object or document
- Increasing trouble with planning or organizing

Top Ten Early Signs of Alzheimer's

1. Memory loss and forgetting things that just happened
2. Changes in mood and personality (this can be very subtle, such as someone who is strong-willed by nature but grows increasingly stubborn)
3. Social withdrawal
4. Misplacing important things
5. Difficulty completing familiar tasks
6. Confusion of time and place
7. Poor judgment and decision making
8. Struggling to communicate
9. Changes in vision
10. Inability to plan or solve problems

Middle Stage: Moderate Alzheimer's Disease

The middle stage is typically the longest one and can last for many years. As the disease progresses and symptoms become more pronounced, those with Alzheimer's will require a greater level of care. Although they may still remember significant details about their life, they may have greater difficulty performing tasks, such as paying bills and tending to household chores.

A witness to someone in this stage may notice the person confusing words, getting frustrated or angry without provocation, or acting in unexpected ways, such as refusing to bathe or dress appropriately. Damage to nerve cells in the brain can make it challenging to express thoughts and perform routine daily tasks. At this point, symptoms will be noticeable to others and may include:

- Forgetting events or part of one's own personal history
- Feeling moody or withdrawn, especially in socially or mentally challenging situations
- Being unable to recall one's own address or telephone number or the high school or college from which one graduated
- Being confused about where one is or what day it is
- Needing help choosing appropriate clothing for the day or an event
- Having trouble controlling bladder and bowels
- Changes in sleep patterns, such as sleeping during the day and becoming restless at night
- Wandering and becoming lost
- Personality and behavioral changes, including suspiciousness and delusions or compulsive, repetitive behavior like hand-wringing, repeating comments, or repeating the same gesture

Late Stage: Severe Alzheimer's Disease

Dementia symptoms are serious in the final stage. People lose the ability to respond to their environment, carry on a conversation, and, eventually, control their movements. They may still say words or phrases, but communicating in general, including sensations of pain, becomes difficult. As their memory and cognitive skills continue to decline, very noticeable personality changes occur and individuals need extensive help with daily activities. At this stage, a person may:

- Need around-the-clock assistance with daily activities and personal care
- Lose awareness of recent happenings, as well as of their surroundings
- Lose basic physical abilities, such as the ability to walk, sit, and, eventually, swallow
- Experience increasing difficulty communicating
- Become vulnerable to infections, especially pneumonia

Surprisingly, no single diagnostic test can determine if a person has Alzheimer's disease. Even if a brain scan to look for beta-amyloid is performed, we don't have definite parameters established for what's normal and not and whether any buildup of amyloid in the brain is truly causing the symptoms. There's disagreement among pathologists as to what exactly constitutes "enough" plaque, and in what locations, to make the diagnosis of Alzheimer's disease. The U.S. Preventive Services Task Force does not recommend screening, while some neurologists do. Health care providers typically do not make a diagnosis of dementia unless the symptoms are already so severe that they interfere with a person's activities of daily living. Adding to the challenge is the fact some doctors, especially primary care physicians, hesitate to diagnose dementia and are often ill-equipped to deliver the news. They sometimes cling to outdated thinking that says there's nothing to be done for those identified as being at greater risk for dementia or who have already been diagnosed. Primary care providers grappling with these issues can benefit from the Gerontological Society of America's four-step KAER process, a tool kit developed to help them detect cognitive impairment and provide earlier diagnosis, which can greatly improve quality of life for their patients.

Arriving at a diagnosis usually entails the help of several specialists such as neurologists, psychologists, geriatricians, and geriatric psychiatrists, plus a variety of approaches and tools. A standard medical workup for Alzheimer's disease often includes structural imaging with magnetic resonance imaging (MRI), or computed tomography (CT). Structural imaging can reveal other causes of the patient's symptoms such as tumors, small or large strokes, damage from severe head trauma, or a buildup of fluid in the brain. A third type of scan, PET, can show patterns of brain activity and whether the amyloid protein is accumulating. But again, these scans have limitations. They are best used within the context of other clinical findings. Most of these tests are not going to detect Alzheimer's; rather, they rule out other conditions that may cause symptoms similar to Alzheimer's but require different treatment.

DEMENTIA MIMICS

Let's take a closer look at some dementia "mimics" because many of these conditions can be treated successfully.

Normal Pressure Hydrocephalus (NPH)

It was a remarkable experience for me as a neurosurgeon to see someone successfully treated for normal pressure hydrocephalus. As with most other patients, the man I was treating had been given a diagnosis of Alzheimer's and had been treated for two years. Having reached out for second and then third opinions, he was finally found to have NPH, a gradual buildup of cerebrospinal fluid (CSF) in the brain, which results in swelling and pressure that can damage brain tissue over time. My patient had classic symptoms of NPH, including walking and balance problems, urinary incontinence, and memory difficulties. When I saw him and reviewed his CT scan, I was pretty confident he was going to benefit from draining the excess fluid. First, I did a lumbar puncture and inserted a lumbar drain to investigate if draining large amounts of CSF would help with his symptoms. The plan was to have the physical and cognitive therapists evaluate him after a couple days to see if he had any improvements.

Amazingly, after the first day, he sat up on his own when I walked into the room. He was so thrilled about his improvement that he almost pulled his lumbar drain out as he started showing off how well he could walk. He told me he felt "uncongested" after having the fluid drained. It was dramatic and emotional for him and his family. They had basically resigned themselves to a life with Alzheimer's.

After that, I placed a shunt (a catheter that drains cerebrospinal fluid from the ventricles in his brain, and diverts it to his abdomen) and he continued to improve. It was one of the most satisfying operations I have performed because it's not common to achieve such a quick response to what could otherwise be a serious and untreatable brain problem. Best estimates are that close to a million people have NPH, and fewer than 20 percent are properly diagnosed. Not everyone will have improvement with drainage of CSF, and few will have as dramatic a turnaround as this patient. Still, it is one of the dementia mimics that needs to be evaluated.

Medications

More than half of Americans take at least one prescription drug, and more than half take an average of four.[5] The older we get, the more likely we are to take drugs for various conditions, especially in the United States. Twenty percent of Americans are on five or more prescription medications. These include drugs like antidepressants, antibiotics, statins, opioids, benzodiazepines (used for anxiety and sleep), and blood pressure medications. We often don't ask or think about a drug's side effects or interactions with other drugs when a doctor prescribes them, and rarely do we consider effects that can mimic a disease like Alzheimer's. We just go ahead and take these drugs prescribed for us. But many commonly used drugs can trigger cognitive symptoms. As we age, the body metabolizes and eliminates medication less efficiently, allowing drug levels to build up and cause memory glitches. Which ones? The most likely culprits are some I just named: narcotic painkillers (opiates), benzodiazepines, muscle relaxants used after injury, and steroids.

This is why it's critical to tell your doctor about every drug you take, including supplements and over-the-counter medications. You might assume your doctor already knows everything that's in your medicine cabinet (and perhaps all your prescriptions came from your own doctor), but it helps to make sure you remind and make your doctor aware of additional things you take, including vitamins and dietary supplements that don't require a prescription.

One particular class of drugs has gained notoriety in dementia circles: anticholinergics. As the name implies, an anticholinergic agent is a substance that blocks the neurotransmitter acetylcholine in the central and the peripheral nervous system. Acetylcholine is responsible for transferring signals between certain cells that affect specific bodily functions. In the brain, acetylcholine has a role in learning and memory; elsewhere in the body, it stimulates muscle contractions. Anticholinergic drugs' effects make them a candidate to treat an array of illnesses such as depression and Parkinson's disease, as well as gastrointestinal disorders, urinary incontinence, epilepsy, and allergies. Benadryl—the popular antihistamine many of us have in our medicine cabinets and also find in over-the-counter cold remedies and sleep aids—owes its main ingredient to an anticholinergic: diphenhydramine. But here's what's increasingly a concern: This class of drugs may also increase a patient's risk of dementia by more than 50 percent. And it's estimated that 20 to 50 percent of Americans age sixty-five and older take at least one anticholinergic medication. A 2019 study published in *JAMA Internal Medicine* revealed that in men and women age sixty-five and older, taking an anticholinergic for three or more years was associated with a 54 percent higher risk for dementia than taking the same dose for three months or less.[6] These are not drugs you want to be taking long term to maintain a clear head.

If you take an anticholinergic, talk to your doctor about the risks versus benefits and see if there are other alternative options. We don't know yet what the long-term effects can be from using these drugs. By some measures, researchers have found greater incidences of dementia among patients prescribed anticholinergics for depression, urological problems, and Parkinson's disease than among older adults who were not prescribed these drugs. It is still not clear whether it was the medications that increased the risk or the underlying malady that required the medication, but the new incidences of dementia were found up to twenty years after exposure to the drugs.

Drugs that May Increase Risk of Dementia

- anticholinergic antidepressants (e.g., paroxetine, Paxil)
- antiparkinson drugs and antihistamines (e.g., diphenhydramine, Benadryl)
- antipsychotic drugs (e.g., clozapine, Clozaril)
- drugs for overactive bladder (e.g., oxybutynin, Oxytrol)
- antiepileptic drugs (e.g., carbamazepine, Tegretol)

Depression

This area is tricky. Severe depression can frequently cause symptoms of dementia, sometimes called pseudodementia. When the depression is successfully treated, the cognitive impairment improves. But it is important to know that the individual will still have a higher risk of developing dementia later in life. Complicating matters is the fact that people with various forms of dementia live with a higher risk for depression, largely due to the damage of the emotional circuitry in the brain. You can see the vicious cycle here. This is why it's imperative to have anyone suspected of suffering from dementia also evaluated for depression. A psychiatrist, neurologist, or geriatrician affiliated with a memory disorder clinic or major medical center will already have this evaluation in the protocol.

Multiple studies have shown severe depression in older people with normal memory is linked to developing dementia within a few years. Because we now understand Alzheimer's and related diseases can begin to develop decades before the memory loss symptoms of the disease are obvious, it is unlikely that depression leads to rapid-onset Alzheimer's disease.

Therefore, it's possible that late-life depression is one of the early signs of Alzheimer's disease. Sometimes, it can be difficult to distinguish between certain aspects of depression and mild memory loss in normal aging and what may be a disease. Technology—including cerebrospinal fluid protein levels and amyloid PET imaging—is now available to determine whether changes in feelings and emotions or in memory that come with older age are consistent with Alzheimer's disease. Most clinicians agree depression symptoms should be treated through medications or non-pharmacological approaches regardless of whether Alzheimer's disease is present.

Urinary Tract Infection (UTI)

UTIs are caused by a buildup of bad bacteria in the bladder, ureters, urethra, or kidney that cause infections. They often present differently in older people because they rarely have the typical symptoms of a high fever or pain, especially when urinating. Instead, the person may experience sudden memory problems, confusion, delirium, dizziness, agitation, or even hallucinations. UTI-induced confusion is most likely to occur in people who are older or already have dementia. Eradicating the infection with proper treatment can help ease symptoms.

Vascular Dementia

Vascular dementia can be caused by any number of adverse cardiovascular conditions or events including a massive stroke, which results in losing function in parts of the body or having trouble speaking, or a series of ministrokes. The latter, called subcortical vascular dementia, can result in the person showing signs of cognitive impairment without any idea of having suffered the strokes because they occurred silently. The treatment protocol for this is usually reducing the risk of further strokes through improving diet and exercise, controlling blood pressure, and engaging in cognitive rehabilitation. Sometimes drugs approved to treat Alzheimer's can help. Vascular dementia can also be caused by damaged blood vessels in the brain as a result of diabetes, high blood pressure, or atherosclerosis (hardening of the arteries).

Nutritional Deficiencies

AARP's survey found that over 25 percent of adults in the United States fifty and older are taking a supplement for brain health purposes, which is a big waste of money for most of them. The Global Council on Brain Health's report on supplements does *not* recommend supplementation for brain health unless a health care provider has specifically identified a nutritional deficiency. When nutritional deficiencies do exist, however, they can lead to symptoms of dementia because of their influence on the metabolism and downstream effects. The most common deficiencies include vitamin B12, niacin (the deficiency causes a disease called pellagra), and a general imbalance of nutrients called protein calorie malnutrition due to lack of overall intake of healthy foods. Luckily, deficiencies are rare in the Western world and can often be remedied through diet and supplementation.

Underlying Infection

As noted earlier, infections can trigger symptoms of dementia. Syphilis, for example, has long been associated with increased risk for dementia due to its effects on the nervous system and brain. Research is currently underway to understand other infections that can have brain-wasting consequences—from Lyme disease to other vector-borne ailments such as bartonellosis, which is caused by the Bartonella bacteria.

Brain Tumor

Having a benign brain tumor called a meningioma can sound dire, but it can be better than getting diagnosed with dementia. Several of these tumors can be surgically removed, unlike the plaques causing Alzheimer's dementia. These tumors can press on certain parts of the brain and cause cognitive dysfunction. The key here is early assessment so these tumors can be removed in their early stages, which can increase the chances that cognitive changes can be reversed. Otherwise, the longer they remain and grow, the harder they are to remove, and there's much greater risk of permanent damage.

Subdural Hematoma from a Head Injury

A subdural hematoma occurs when abnormal bleeding (usually caused by an injury) leads to blood collecting between the dura (the outermost of the meninges, which are layers of tissue that surround the brain) and the brain. A buildup of pressure caused by the hematoma can lead to dementia-like symptoms. These can be relatively easy to drain surgically, especially if the hematoma has become liquefied. Small ones may even go away on their own over time. Because these blood collections may take some time to accumulate, patients may forget the minor head injury that often causes them. Something as seemingly innocuous as bumping your head while getting into your car can show up as a subdural hematoma days or weeks later, especially in someone who is elderly.

Traumatic brain injuries in general can cause memory loss that mimics symptoms of dementia, especially when the injury happens in regions of the brain associated with learning and emotions. In 2019, researchers at UCLA and the University of Washington revealed that MRI scans in development today will help doctors in the future distinguish between Alzheimer's disease and traumatic brain injury.[7] It's important to be able to make this distinction because it can inform the proper treatment. Note that the risk of falling in general tends to increase the older we get, so preventing falls to begin with will go a long way to avoid breaking bones and suffering from a traumatic brain injury.

Alcohol Misuse

Alcohol dementia (or alcohol-related dementia) is caused by long-term, excessive consumption of alcoholic beverages and is increasingly a concern for doctors because the amount of drinking has been on the rise in our society, especially among older folks. In addition to destroying brain cells in areas critical for memory, thinking, decision making, and balance, heavy drinking can also lead to injury and increase the risk of other health problems that can impair cognitive function (such as liver damage). Certain medications combined with alcohol can also cause memory issues and other side effects. The effects of alcohol abuse can sometimes be reversed, but the first step is to abstain from alcohol, which can be difficult for long-term drinkers.

THE MEDICAL WORKUP

Anyone concerned about the possibility of having a form of dementia should be seen as soon as possible for a complete medical workup. This should include a review of the following:

- One's medical history and a complete physical with lab work (blood and urine)
- Psychiatric history and history of cognitive and behavioral changes
- Current and past illnesses
- Medications and supplements
- Medical conditions affecting other family members
- Lifestyle habits like diet, exercise, and use of alcohol

The combination of the physical exam and laboratory **tests can help spot** health issues that can cause symptoms of dementia, **such as depression,** untreated sleep apnea, side effects of medications, **thyroid problems,** certain vitamin deficiencies, and excessive alcohol consumption. Even hearing loss can be a warning sign; although we don't fully understand the connection, new research points to moderate and severe hearing loss being a significant risk factor for dementia. The good news is that for some people, treating the hearing loss may prevent or delay progression of the disease.[8]

The neurological part of the exam can include a brain imaging study and assessments that can gauge a person's range of everyday mental skills. For example, is the individual aware of symptoms? Does she know the date, time, and where she is? Can he remember a short list of words, follow instructions, and do simple calculations? Among the commonly used tests to identify potential problems are these:

- *The Alzheimer's Disease Assessment Scale–Cognitive Subscale (ADAS-Cog)* is one of the more comprehensive and widely used tests. Researchers often use it in cognition research and clinical drug trials for antidementia medications. It was developed in the 1980s and primarily measures memory, language, and orientation (e.g., how someone solves a problem). A noncognitive part measures things like mood, attention, and motor activity, but this is not used nearly as much as the ADAS-Cog section, which can be done on paper or electronically. Unlike some of the other tests that take minutes to complete, the ADAS-Cog takes about thirty to thirty-five minutes and consists of eleven sections led by a test administrator who adds up points for errors in each task. The greater the total score (out of 70), the greater the dysfunction. Research shows that a normal score for someone who does not have Alzheimer's or another type of dementia is 5. Studies also show that 31.2 is an average score for people who have been diagnosed with probable Alzheimer's disease or mild cognitive impairment, though critics charge that the ADAS-Cog is not that effective in rating the severity of that impairment and mild cases of dementia. It is, however, considered better than many other tests.

- *The Mini-Mental State Exam,* also called the Folstein test, is a simple questionnaire that takes about ten minutes to complete. Dating back to 1975, it's also one of the most commonly used basic screenings for dementia in clinical settings and has a maximum score of 30 points. It evaluates attention and calculation, recall, language, ability to follow simple commands, and orientation (time and place). It can be done on paper and requires no fancy equipment or even a computer. A score of 20 to 24 suggests mild dementia, 13 to 20 suggests moderate dementia, and less than 12 indicates severe dementia. On average, the MMSE score of a person with Alzheimer's declines about two to four points each year.

- *The Mini-Cog test* is even simpler and briefer than the MMSE. It takes just three minutes to complete and has two components: a three-item recall test for memory and a clock-drawing test whereby the person is asked to draw the face of a clock showing all twelve numbers in the right places and a time specified by the examiner.

- *The Self-Administered Gerocognitive Examination (SAGE),* another simple test done on paper, was developed at the Ohio State University's Center for Cognitive and Memory Disorders. Like the other tests, this one asks fundamental questions to show how well the brain is working, including language, memory, and problem solving. It takes about fifteen minutes, and although it's marketed to be done at home or in a doctor's office, I recommend that anyone looking to use these tests do so in a formal setting under the guidance of a qualified doctor if possible.

Many other cognitive exams are available. In research circles, multiple different assessments are often used because no single test is a total diagnostic. In other words, these tests alone do not diagnose dementia. They are *assessments*—they evaluate general cognition and measure the amount or degree of impairment. The results become part of the entire medical workup to determine whether someone is diagnosed with a form of dementia.

Computerized cognitive tests are becoming increasingly popular among physicians, and they can have advantages over the older, written exams. They can be more precise in their evaluations of thinking, learning, and memory, as well as administered exactly the same way in the future to document changes. Using both clinical tests and computer-based tests can give physicians a clearer understanding of cognitive difficulties that patients are experiencing. The FDA has cleared several computerized cognitive testing devices for marketing, including Automated Neuropsychological Assessment Metrics (ANAM), CANTAB Mobile, COGNIGRAM, Cognision, and Cognivue.

It's important to note that all of these tests—written questionnaires or computerized—should be administered by a professional familiar with their interpretation. As simple as some of these tests are, don't attempt to self-evaluate using a test you can download or take online. I also recommend that you don't "cheat" by studying these kinds of tests over the Internet before being tested in a professional setting. These tests aren't perfect, and they can be gamed. Remember that the goal is to get a clean and unbiased screening assessment. It's also important to note that current testing is not 100 percent accurate 100 percent of the time, so it's helpful to get second and even third opinions when possible.

Do you have to take the assessment at a top facility? It's a question I get regularly. The answer is "not necessarily," but be sure that you're with a doctor and a team that regularly sees and diagnoses all forms of dementia. As a starter, if you are older, finding a good geriatrician is an excellent idea for every patient. There is a chronic shortage of geriatricians across the country, particularly now as we have growing numbers of older people. If you can't find one, make sure your primary care provider has some experience. You don't want to be diagnosed by someone who rarely deals with dementia and who will not have recommendations for you going forward. Also remember that the people who notice the earliest signs are often family members, coworkers, and friends, not the individual beginning to show signs of cognitive decline or even a doctor. The details noted by family members are very important and can be crucial to figuring out the time line of the disease, the speed of progression, and whether there could be some other cause. After that, a medical team—typically a neurologist, psychiatrist, and psychologist—would be ideal to help assess this disease.

NATIONWIDE PROGRAMS: WHERE TO FIND HELP

The Alzheimer's Association is a leading voluntary health organization in Alzheimer's care, support, and research. It delivers education, support, and services for people diagnosed with Alzheimer's, their families, caregivers, health care professionals, and the general public. The organization maintains a 24/7 Helpline that's free and confidential: 800-272-3900. Call anytime to receive reliable information, advice, and support. Trained and knowledgeable staff are ready to listen and can help you with referrals to local community programs and services, dementia-related education, crisis assistance, and emotional support. Care consultations are provided by specialists and clinicians.

The Alzheimer's Association may be the oldest organization focusing on this disease, but it's certainly not the only one. In fact, there are many local organizations that are not part of the Alzheimer's Association that do great work and provide a wealth of resources. A list of some of the finest institutions for diagnosing, treating, and researching dementia and, in some instances, other brain-related ailments like Parkinson's and stroke follows. This is not an exhaustive list, so don't hesitate to check out places near you that a friend or your doctor can vouch for. In addition, the National Institute on Aging keeps an online directory of all the research centers it collaborates with across the nation.

- *AARP* maintains a comprehensive library of resources for people living with dementia (aarp.org/disruptdementia) as well as for their caregivers (aarp.org/caregiving) where, by answering three quick questions, caregivers can get personalized information and resources based on their specific concerns. Partnering with the Alzheimer's Association, AARP also has a community resource finder for social and medical services, housing, and programs that can be your link to local support services. And don't miss AARP Staying Sharp, a holistic program that helps you take control of your brain as you age, designed to help you keep your brain sharp. Visit www.stayingsharp.org/keepsharp to get started.

- *The Cleveland Clinic's Lou Ruvo Center for Brain Health* provides diagnosis and ongoing treatment for patients with cognitive disorders and support services for family members who care for them, integrating research and education at every level. The center offers services in Cleveland and Lakewood, Ohio; Las Vegas, Nevada; and Weston, Florida. my.clevelandclinic.org/departments/neu rological/depts/brain-health.

- *The Dementia Action Alliance* is a national advocacy and education organization of people living with dementia. This is an extraordinary organization that aims to take the stigma and misperceptions out of the disease and give people the tools to live courageously and purposefully with the disability. When I spoke with some of the members of its board of directors, I was moved by their perspective on how to speak about dementia and communicate with patients using positive, hopeful terms. They use the word *carepartners* in place of caregivers, and highlight that it's not about surviving the disease—rather, it's about *thriving*. Some community members feel that life is actually better after their diagnosis because it opens doors and creates new opportunities. *See me, not my dementia* is one of their mottos. Check out the alliance's site and the resources available online at daanow.org. It publishes two small handbooks: one for the person diagnosed, and the other for family and friends.

- *The Family Caregiver Alliance* has been around for more than forty years, but it's poised to expand its reaches and visibility as it promotes new programs that help connect institutions and health care providers with its evidence-based Best Practice Caregiving solutions for high-quality dementia care. It supports a searchable database of effective programs for family caregivers of persons with dementia. The goal is to increase knowledge and adoption of non-drug, evidence-based programs for family and friend caregivers by health care and community service organizations. It ultimately helps individuals and families gather the information they need and find programs in their area that can help with managing a diagnosis. www.caregiver.org.

- *The Mayo Clinic's Alzheimer's Disease Research Center* offers opportunities to participate in drug trials, clinical research projects, special programs, support groups, and education events. It has bases in Scottsdale, Arizona; Jacksonville, Florida; and Rochester, Minnesota. You can request an appointment online just by filling out a form. www.mayoclinic.org.

- *The Memory Disorders Program at New York–Presbyterian/ Weill Cornell Medical Center* has set the standard of care for the management of memory disorders. Program physicians coauthored the American Academy of Neurology guidelines for the diagnosis and treatment of dementia and hydrocephalus and for the use of genetic testing in families with Alzheimer's disease. In 2013, the center established the Alzheimer's Prevention and Treatment Program where people interested in lowering their risk for Alzheimer's can be followed over time and receive a personalized plan of care based on their risk factors, genes, and past and present medical conditions. weillcornell.org/services/neu rology/alzheimers-disease-memory-disorders-program/about -the-program.

- *The National Institute on Aging funds Alzheimer's Disease Research Centers (ADRCs)* at major medical institutions across the United States. Researchers at these centers are working to translate research advances into improved diagnosis and care for people with Alzheimer's disease, as well as finding a way to cure and possibly prevent Alzheimer's. www.nia.nih.gov/health/alzheimers-disease-research-centers.

- *UCLA's Alzheimer's and Dementia Care Program* helps patients coordinate their care between their primary care physician and a dementia care specialist who has a nursing background. Patients and their families develop personalized plans based on individual needs, resources, and goals. dementia@mednet.ucla.edu, www.uclahealth.org/dementia. Also see UCLA's Longevity Center: www.semel.ucla.edu/longevity/.

In Their Words

Brian Van Buren is a hero in the dementia community and someone who is not about to give up. Diagnosed in 2015 with early-onset Alzheimer's at the age of sixty-four, he's living with the disease and has become an outspoken advocate for the African-American and LGBTQ communities—places where dementia is sadly stigmatized and hushed. I was charmed by his candor and humor when I spoke with him. "I came out in the 1970s not knowing I'd have to come out again when I was diagnosed with dementia," he candidly told me. Brian wears a button every day that says "Living with Dementia," and it's quite the conversation starter. He's on the board at the Dementia Action Alliance and is a frequent speaker at large-scale events and on radio shows. Life does not end after the diagnosis, he told me. It doesn't have to be a death sentence and the idea that you should "go home and get your affairs in order" is not the view to focus on—at least not initially. You will go through a grieving process at first and then carry on with knowledge about what to do. Brian stays centered with the help of a life coach who specializes in dementia and he participates in what's called couch surfing, a homestay program. People from around the world visit him and have a place to stay for a few nights for free. He's enjoyed the company of more than one hundred guests and will continue to host for as long as he can.

THE FUTURE

There's a lot that can be done to delay the progression of the illness. I can't repeat this enough: Early detection is critical. You may be wondering why this is so important given the lack of effective drugs or a cure. I have found that it can be reassuring to family members when a loved one is diagnosed, even if it is Alzheimer's disease, because it is finally an answer at the end of an often long and confusing journey. It enables people living with dementia to participate in their care plan and express their views on what they want and need before it becomes too difficult for them to communicate with their health care providers and loved ones. It also allows planning for the future, including logistics and the cost of care. Early diagnosis may also make someone more eligible for certain clinical trials, which is critical for future effective treatments. The goal should be to *enable* the person with dementia—not disable them. People with dementia still have a lot to offer and can continue to learn new things. It is sometimes possible for people to live twenty years or so after the first symptoms appear. Rates of progression differ widely, and in the future, people will realize that we can manage the symptoms so people can live as well as possible for as long as possible. People living with dementia can do a lot to improve their quality of life. Again, engaging them in the care planning process is critical to enable health care providers to deliver person-centered care that can dramatically improve health outcomes and the quality of care.

Just a few decades ago, nobody wanted to talk about cancer; today cancer patients take pride in talking about their illness and forging ahead with hope and resolve. We've destigmatized cancer and developed strategies for treating each cancer patient individually based on their particular cancer, values, resources, and family dynamics. We are on the cusp of revolutionizing how we view and treat dementia, improving people's experience with it—from the patients to the caregivers. There is also a lot that people living with dementia can do to improve their quality of life and delay the progression of the illness. Engaging them in the care-planning process is critical to enable health care providers to deliver person-centered care that can dramatically improve health outcomes and the quality of care.

It's estimated that delaying the onset of dementia by only five years can cut the incidence rate in half, vastly improving life and well-being for people and reducing health care costs for families and society. Over the next few years, I believe there will be significant progress in early detection techniques for Alzheimer's with the help of technologies like artificial intelligence and big data mining to find biomarkers. Such biomarkers can range from the usual suspects like certain lab tests to novel findings like losing one's sense of smell. New research suggests that having a weak sense of smell could be an early warning sign of cognitive decline. The neurodegeneration that occurs with these ailments affects brain circuits linked to our olfactory system. Testing someone's sense of smell using common scents—clove, leather, lilac, smoke, soap, grape, lemon—is inexpensive and noninvasive and may lead to new therapies.

Blood tests for dementia may come sooner than previously thought—even within the next few years. Scientists are inching closer to such tests that can help screen people for possible signs hidden in the blood that may not come with outward signs of a problem. A blood test is a lot more economical and easier to perform than other tools that involve brain scans and spinal fluid tests. If you can know a potential diagnosis years ahead of developing symptoms of a brain disorder, that could change your brain's future with interventions you can execute right away.

Q: Should I take a dementia screening test at home today that I can buy online or download? And what about getting a brain scan?

A: A number of these tests have come on the market targeting consumers and they don't require a doctor to prescribe or even oversee. None of these tests have been scientifically proven to be accurate and should be met with caution. The last thing you want to get is a false-positive, meaning the results say you have dementia when you do not. False-positives are highly unlikely when you visit a physician to seek a potential diagnosis. Avoid these kinds of tests, even if they are tempting. Anyone should be evaluated within the context of an ongoing relationship with a health care professional.

As far as PET (positron-emission tomography) brain scans go, again you may want to hold off on paying for such tests. Not only are these expensive (neither Medicare nor private insurers cover their costs, which can be between $5,000 and $7,000), but they can have unintended consequences. Positive scans that detect amyloid plaques do not mean you will go on to develop dementia, yet they can lead to costly and ineffective treatments. Negative scans do not mean you won't develop the disease. Interestingly, biostatisticians at UCLA have calculated that a seventy-five-year-old man with amyloid has a little more than a 17 percent lifetime risk of developing Alzheimer's dementia; for a woman that age, her changes are about 24 percent with a longer life expectancy.[9] Until these tests become more reliable and useful, leave them for the researchers using them in clinical laboratory settings.

TREATMENTS: DRUG-BASED AND PEOPLE-BASED

The complexity of dementia makes it uniquely difficult to treat, more so than just about anything else in the world of neuroscience. We have little in our arsenal to combat the disease once it has taken hold and begun its march forward. The two FDA-approved classes of drugs to lessen symptoms of Alzheimer's disease aim to keep brain cells communicating with one another so the brain can function normally, but these medications are far from a promising therapy and come with their own side effects. They can temporarily improve symptoms of memory loss and problems with thinking and reasoning, but they lose their effectiveness as the disease progresses. In other words, these treatments don't stop the underlying decline and death of brain cells; they just throw some obstacles in the way to buy time.

The first class of drugs encompasses cholinesterase inhibitors, which work by inhibiting the breakdown of acetylcholine and keeping it at healthy levels. Acetylcholine, you may recall, is an important neurotransmitter in the brain, responsible for sending signals in the nervous system and plays a key role in memory. (By contrast, anticholinergics block the action of acetylcholine. So, to be clear, cholinesterase inhibitors and anticholinergics having opposing effects in the body.) In clinical trials, cholinesterase inhibitors show modest effects against the functional and cognitive decline of people with Alzheimer's disease. You might know these medications by their more common brand names: Aricept, Exelon, and Razadyne. Acetylcholine breaks down naturally in everyone, but the process is much worse in people with Alzheimer's, who have low levels of acetylcholine in their brain.

The second type of treatment is an NMDA receptor antagonist that also works primarily by keeping lines of communication open between brain cells. The drug, memantine (Namenda), regulates the activity of glutamate, another chemical messenger involved in brain functions such as learning and memory. Glutamate is critical because when brain cells are damaged by Alzheimer's disease, they pump out too much of it, which damages even more brain cells.

These two types of drugs often are prescribed together, especially in later stages of the disease. Other drugs can also be prescribed to treat symptoms from other conditions based on an individual's diagnosis. Someone with mood disorders and sleep disturbance on top of a form of dementia may benefit from additional medications, for example. The tricky part, of course, is knowing which drugs to use in combination without a worsening of side effects or a canceling-out effect. People with Parkinson's disease, for instance, may benefit from an anticholinergic for controlling tremors, but not at the risk of speeding up Alzheimer's disease. There is growing concern that if someone is taking both types of drugs—cholinesterase inhibitors and anticholinergic medications—they will antagonize each other, and neither will work.

In 2018, the FDA announced revised guidelines for neurological disorders that will make clinical drug trials for preclinical Alzheimer's easier to conduct. This represents a major policy shift that the FDA hopes will lead to better treatment at the earliest stages of the disease, where medical intervention is most promising. Such trials should also lead to better treatments that stop or delay the disease process before it progresses.

Another hopeful note is the Coalition Against Major Diseases, an alliance of pharmaceutical companies, nonprofit foundations, and government advisers that has forged a partnership to share data from Alzheimer's clinical trials. It has also collaborated with the Clinical Data Interchange Standards Consortium to create data standards. Data sharing can speed research and drug development. As you read this, researchers are hard at work trying to come up with effective therapies. Until we have reliable solutions, one thing top scientists agree on is that when a diagnosis comes in and someone is suffering from a neurodegenerative condition, you don't give up. Like Sandy Halperin, you can become a voice, an advocate, and a model patient.

It's important to understand that "treatment" may not come in the form of a super drug. Treatment can be the quality of care and the lifestyle plan that's set upon diagnosis. How a person is cared for by a loved one—the person who is the guide or shepherd through the process—is critical to how a patient progresses. Effective interventions to improve quality of life are growing, though they need to be radically accelerated. "High-quality dementia care" may sound like an oxymoron, but it doesn't have to be—especially with the aid of the Internet that can connect people around the world and build communities like those supported by Dementia Friendly America, the Dementia Action Alliance, and other organizations. There's a movement away from the "nothing can be done" mentality that has unfortunately stained this area of medicine for far too long and, frankly, set it back. Katie Maslow is excited about the prospects of these new programs. As a former scholar-in-residence at the Institute of Medicine, veteran policy-related researcher for the Alzheimer's Association, and now visiting scholar at the Gerontological Society of America, she knows a thing or two about best practices for managing dementia. She echoes what other experts have told me: Each patient must be treated individually because everyone is different. What works for one person might not help another. The drumbeat message of "search for the cure" eclipses other areas where we should be paying attention—places where we can proactively keep people in stable, early stages of the disease and improve their experience and quality of life.

Dr. David Reuben is a gerontologist at UCLA who holds many credentials. In addition to his role as Chief of the Division of Geriatrics Medicine and Professor at the David Geffen School of Medicine at UCLA, he maintains a clinical primary care practice and also directs the UCLA Claude D. Pepper Older Americans Independence Center and the UCLA Alzheimer's and Dementia Care program, which I mentioned in the list of programs where you can find help. Like all the other experts I spoke with, Dr. Reuben emphasizes the importance of an individualized approach to taking care of patients with dementia and focusing on "the dyad"—the patient and the caregiver. Cookie-cutter approaches are not going to work; tailoring the intervention to a patient's disease, personal resources, and goals are what leads to better outcomes and higher quality of life. And even though many caregivers find the job gratifying, that doesn't mean it's stress-free. As we'll see in this next chapter, managing the primary caregiver's health is just as important as managing the person with dementia. There will be a lot of twists and turns and no one can prepare for it all. According to Dr. Reuben, when it comes to a person's experience with dementia, the most important person is not the doctor—it's the caregiver.

CHAPTER 11

Navigating the Path Forward Financially and Emotionally, with a Special Note to Caregivers

From caring comes courage.

LAO TZU

While working on this book, I was struck by how challenging it is for families to navigate the best way to care for their newly diagnosed loved one. Sadly, I realized that some families often stopped talking about their family members with cognitive impairment and became conflicted by the idea of putting the person into an extended-care facility. Concerns about affording such care are mixed with worries about the quality of care they'd receive. The average cost for a semiprivate room in a nursing home in the United States is over $7,000 per month and about $8,000 for a private room.[1] For people with severe memory issues who need extra care and attention, the costs are even higher. One-bedroom units in assisted living facilities can be a little less expensive, but with fewer staff and skimpier training they may be less than ideal, especially for people with Alzheimer's disease or related dementias. There are superb facilities and excellent staff providing high-quality care to people living with dementia in long-term care facilities every day all across the country. But even if you can afford the prices, there are significant problems in many long-term facilities. In recent years, I have reported on bad apples in the assisted living industry, which is sorely underregulated, leaving some residents in unsafe environments with inadequate care. Worse, some residents are mistreated and abused. This includes facilities that are marketed as specializing in memory or dementia care. Construction of memory care units in assisted living facilities is the fastest-growing segment of senior care. For all of these reasons, the place I will tell you about in the next section is perhaps the most extraordinary of all my journeys.

As someone who has traveled to more than a hundred countries around the world, there is a question I often get: Which of all those places has been the most remarkable and why? My mind quickly scans through war zones, natural disasters, outbreaks, and other scenes where I have witnessed terrible human suffering soon followed by heroic stories of people rising to the situation and doing their new life's work in incredible ways. For them, it is true that necessity is the mother of invention, and the stories of families dealing with dementia are no different.

IT TAKES A VILLAGE

Within the city of Weesp, just minutes outside Netherlands' capital city of Amsterdam, is a gated model village known as De Hogeweyk (a "weyk" refers to a group of houses similar to a village). It was first described to me as a place where a grand experiment has been taking place for over a decade that could fundamentally change the way patients with advanced dementia live out the rest of their lives. The media are rarely allowed inside. I was fortunate to be invited by the founders a few years ago to see it for myself.

The idea for the facility started when two Dutch women, who had both worked in traditional extended care facilities themselves, had a candid conversation about the prospect of their own parents developing dementia and being placed in a conventional nursing home care setting. They thought how jarring it would be to not only lose your memory but to also lose your sense of home and place at the same time. After all, a traditional nursing home is a completely foreign setting with nothing to help ground patients and allow them to have roots. This line of thinking brought them an idea. Their audacious goal was to normalize extended care facilities so that the people living there could experience their lives in a way that felt easy and familiar. The result was Hogeweyk, which was primarily funded by the Dutch government for slightly more than $25 million. This four-acre gated community, which opened in 2009, has been dubbed "Dementia Village," but that sounds a lot worse than what I'm about to describe. Try and visualize it in the following few paragraphs. Let it capture your imagination, as it did mine.

The first thing I noticed was that there is only one way in and one way out. A single pair of sliding glass doors separates Hogeweyk from the outside world, and this is the only place you will find security guards. As you walk into this beautiful Netherlands village, you see the famous Dutch tulips wrapping around bubbling fountains. It feels a bit like a beautiful midwestern college campus with its own amalgamation of streets, squares, dormitories, cafés, street musicians, and theaters. But while college campuses cater to young students, Hogeweyk is strategically designed to meet the needs of those with profound memory loss in their final years of life. And to do that, this facility is very much created to look like the outside world, including restaurants and salons.

Each of the twenty-three two-story dormitory-style homes is stylized to resemble different lifestyle categories to match people's interests and backgrounds. For those who come from the upper classes (*Goois*), for example, there is a housing option with an aristocratic Dutch feel to it in its decor and amenities; the residents often like to attend classical concerts and enjoy high tea. Other choices include housing for people who share Indonesian ancestry or whose religious practices are a priority and who attend services regularly. Residents who had once worked in professional disciplines such as engineering, medicine, or law are grouped together in a unit. The same is true for people who were once artists or carpenters or plumbers. The aim is to place residents in environments where they can live close to people who are likely to have similar past experiences. Each household of six to seven residents manages itself, including cooking and washing, with a staff. The caregivers and house attendants, who collectively outnumber the residents two to one, even use an in-house currency to help their residents "buy" groceries at the fully functional supermarket (though no real currency is exchanged in the village; everything is included).

Outside, there are plenty of gardens and communal places where people are encouraged to move, congregate, and embrace the outdoors rather than stay in their rooms. The focus here is on what residents can do rather than not do, and this place has become a pioneering model for specialized elder care. Experts in elder care have been coming here from around the world to get a glimpse of what growing old with an ailing brain can look like in a vibrant community rather than a depressing, isolating, and lifeless institution. Life is alive here with a variety of social clubs and events, bingo nights, theater events, and even a pub.

As normal as it looks, subtle reminders are everywhere about the tremendous level of planning required to care for an entire village of residents with serious cognitive decline. For example, because wandering is a significant concern, the village is highly secure, with cameras monitoring residents every hour of every day. The elevators are controlled by motion sensors and when someone gets in, it automatically takes that person to the next level. Everyone who works in the village, including the barbers, waiters in restaurants, and clerks at the grocery store or post office, is a trained medical professional—geriatric nurses and specialists—whose primary mission is to provide care far beyond what's typically found in a traditional medical facility. That's what sets this place apart from the average nursing home, with its nondescript buildings, anonymous wards, lots of white coats, nonstop television, and plenty of sedating medications. Here, there are no wards, long hallways, or corridors. The intention is to give people a sense of intimacy even if they no longer have an understanding of what's happening around them or in the world itself. Friends and family are encouraged to visit, and people who live in Hogeweyk's surrounding neighborhoods are all welcome to come in and enjoy some of the facilities, such as the café-restaurant, bar, and theater. This is an important goal because all too often, friends and family members fade away when someone is diagnosed with dementia. It can be an isolating disease and the isolation itself can worsen the prognosis. Keeping patients engaged and socially active is important.

The residents may not necessarily know where they are, but they feel that they are at home—and that is precisely the idea. At Hogeweyk, if someone reaches that single doorway to the outside, a staff member will often say that it's broken. I watched as residents simply turned around and walked back in the other direction. No one is trying to "escape," staff tell me; "they are merely confused." Over time, the residents of Hogeweyk consume fewer sedating medications, have a better appetite, appear more joyful, and live longer than those in standard elder care facilities.

I know what you're thinking: This is straight out of the movie *The Truman Show* where a man played by Jim Carrey discovers his entire life is a TV program. Everything he thinks is real is a mirage, created by television producers. So I had to ask the cofounder of Hogeweyk, Yvonne van Amerongen, if this setup is somehow fooling or duping its residents. She was quick to respond: "Why should they feel they are fooled? We have a society here. . . . We want to help people enjoy life and feel that they are welcome here on this earth." It was one of the most humane things I had heard, allowing people to retain their dignity even as the end is near. Yvonne recalled that when her father had died suddenly of a heart attack several years prior, one of the first things that went through her mind was, *Thank God he never had to be in a nursing home.* That became part of her inspiration for Hogeweyk.

When people move to Hogeweyk, their families know this will be their last stop. Residents will be watched over and comforted until they die, which is typically about three to three and a half years after they enter. Only then does a spot open up for a new resident to enter the village. The Dutch health care system makes Hogeweyk possible; it receives the same funding as any other nursing home in the country. (The cost of care is nearly $8,000 per month, but the Dutch government subsidizes residents to varying degrees. Everyone gets a private room, and the amount each family pays is based on income, but it never exceeds $3,600. It has been operating at full capacity since it opened.)

The staff at Hogeweyk count on the different ways dementia touches the brain to keep that brain engaged. For instance, the part of the brain that equips us with musical talents, including recalling words and placing them to a tune, functions the longest. One couple I met and spent a lot of time with was Ben and Ada. Throughout their marriage of sixty-odd years, they enjoyed making music together as a pastime. Ada would play the piano and Ben would sing. But since Ben developed Alzheimer's disease, their communication inevitably began to falter. Eventually Ben could no longer carry on a conversation. Now a resident of Hogeweyk, Ben relies on music to connect with his wife. I watched as Ada played the piano, and Ben, who came across as an incredibly quiet man when I first met him, suddenly started singing along to some traditional Dutch music. It was a beautiful, magical thing to see, and it helped cushion the blow Ada felt when she left Ben at the end of each day. In her words, "We can't talk anymore about everything, but with singing . . . you can make a good concert together. For me, that's very important."

One of the most important lessons I learned at Hogeweyk is to resist the urge to correct someone with dementia. The toughest conversation I had while visiting was with a resident named Jo. At nearly ninety years old, she was charming and animated, with a smile that warmed the entire room. But she still thought she held a daily job and yet couldn't recall what that job was. "Tomorrow," she said to me, "I'll know it, and I'll have to go to it." She also thought that her parents were still alive and that she'd seen them the day before. When I turned to the resident social worker for help in responding to Jo, the social worker told me that how one responds to such confusion depends on the phase of dementia. In the earlier phases, you can challenge them with a question like "Well, how old are you?" And if they say "I'm eighty-four," you respond, "How old would your parents be?" The person might figure it out and say, "Oh, that doesn't make sense." But what you never want to do is correct people with dementia. If they are asking for dinner, for instance, and they've just had dinner but don't recall the experience, you don't deny them. Instead you might ask if they are hungry without forcing them to recall an experience that is no longer retrievable in their brain.

I noticed a lot of hand-holding between couples, one of whom is fading while the other is holding on. One pair I met, Corrie and Theo, seemed to communicate through their hand-holding. Theo, the healthier of the two, told me that Corrie squeezes his hand whenever she sees something or feels something familiar. They spend the whole day in a clasp, and according to Theo, his marriage is the best it's been in nearly sixty years.

I left Hogeweyk wondering: Could this work in other parts of the world? What would this look like in the United States?

BRACE YOURSELF

The majority of people with dementia in the United States live in their home, and for approximately 75 percent of these individuals, family and friends provide their care.[2] The largest proportion of those caregivers are spouses, followed by children and children-in-law, mostly women. The typical profile of a dementia caregiver is a middle-aged or older female child or spouse of the person with dementia. At least 60 percent of unpaid caregivers are wives, daughters, daughters-in-law, granddaughters, and other female relatives. All told, roughly 60 million Americans are caring for someone with Alzheimer's disease. That's more than twice the number of people living in Texas.

Maria Shriver told it to me straight when I spoke with her about dealing with a loved one's future once the diagnosis is in: "Brace yourself. Take care of yourself. I see a lot of women who have kids and are taking care of a parent, too. They are stressed out, desperate, crying. You have to talk to other family members and get help. Alzheimer's is such an emotional, and financial, and physical journey. No one can do this alone." Maria has walked this path before; her father, Sargent Shriver, was diagnosed with the disease in 2003 when she knew little about it. She helped him get through the disease's process until he died eight years later. The experience galvanized her into becoming one of the world's most vocal advocates for research into not just Alzheimer's disease but brain health, with a focus on women. She has since established the Women's Alzheimer's Movement and has championed many brain-health projects—from award-winning documentaries to collaborations with top scientists—to spread knowledge about the disease's challenges and to lend support to families. She called me back right away when I left word for her that I was writing this book. "Anyone with a brain needs to be thinking about the possibility of Alzheimer's disease," she started off saying, going on to stress the importance of prevention and delay. And as she so often does, Maria pointed out something I hadn't considered before: a glaring contradiction when it comes to Alzheimer's disease in the United States. While women are most likely to be the caregivers, they are also much more likely to develop Alzheimer's disease themselves: Almost two-thirds of Americans with Alzheimer's are women, and a woman's estimated lifetime risk of developing Alzheimer's at age sixty-five is 1 in 6 (compared to 1 in 11 for breast cancer).[3] On top of that, there is a medical research gender gap, meaning that women are less likely to be enrolled in clinical trials despite being significantly more affected by the disease.

For a long time, it was incorrectly thought that women developed Alzheimer's more often than their male counterparts simply because they lived longer. But new research shows a complex set of circumstances to explain this discrepancy between men and women that includes differences in biology as well as how diagnoses are made. For example, given the correlation of early dementia symptoms with perimenopause, researchers have wondered about the protective or destructive effects of estrogen and progesterone. More recently, studies have shown that the telltale tau protein is already more diffusely spread throughout a woman's brain in the early stages of Alzheimer's as compared to a man. That suggests that Alzheimer's disease may affect more areas of the brain in women. From a diagnostic perspective, women tend to perform better on tests of verbal memory during the early and midstages of Alzheimer's, making them more likely to be diagnosed only in the later stages of the disease. There are probably clues to the diagnosis and future treatment of Alzheimer's that lay in these gender differences between men and women, and we haven't explored that nearly enough, as Maria told me. I also spoke to Maria a great deal about the difficulty of caring for a parent as well as children, a reality for so many new caregivers. One thing became clear in my conversations with both experts and people currently immersed in the care of someone with Alzheimer's: Every day feels like a desperate scramble to stay in control.

There is a lack of consistency with treatment plans, coverage, and support. Unfortunately, there are not many Hogeweyk-like communities for dementia patients here, although there may be sometime soon. (The closest thing I found is the Glenner Town Square in Southern California that is an Alzheimer's-friendly facility reminiscent of the 1950s, but it serves only as a day care center. I expect more facilities based on the village-like model and specializing in memory care to crop up in the future.) Most families in the United States struggle to find the right care—and the money to pay for it. The best estimate is that more than 15 million people have someone with Alzheimer's in their family, a number that will continue to grow. The caregivers for someone with Alzheimer's provide an estimated 18.1 billion hours of unpaid care annually. Out-of-pocket costs for Americans with Alzheimer's or other types of dementia are much higher, on average, than for those without. Dementia caregivers spend an average of $10,697 annually out of their own pockets, more than twice the amount spent by caregivers of people without dementia.[4] It is safe to say that advanced dementia is arguably one of the most destabilizing afflictions to a family's emotional and financial health.

I honestly don't know what's worse: the financial or the emotional toll that caring for an individual with dementia costs. If the diagnosis came in for me, I'd immediately worry about my family and their well-being as they try to help me through the illness. I have learned this over the past couple of years of working on this book. The diagnosis is life changing and leads to many immediate questions. What will this mean for me and my family? How do I plan for the future? Where can I get the help I need? How will I pay for it all? Who will be in charge? What happens when I can't make any decisions anymore? Will there be any assets left for my children?

The Alzheimer's Association maintains a wealth of useful information for both people living with Alzheimer's and their caregivers. And it's all free, including a 24/7 hotline: 800-272-3900, and a website: www.alz.org. AARP (www.aarp.org/disruptdementia), and the Global Council on Brain Health (www.aarp.org/gcbh) also maintain lots of free information to answer critical questions about dementia and brain health. AARP maintains a caregiver's hotline from 7:00 a.m. to 11:00 p.m. ET at 1-877-333-5885. The support line is also available in Spanish, at 1-888-971-2013. AARP's online caregiving community is a place where caregivers can join, for free, to talk with other caregivers and get answers from experts in the community. Check out AARP's Caregiving Resource Center (www.aarp.org /caregiving).

Here are a few things I would nail down as soon as possible after a diagnosis. Some are obvious, and others are less intuitive, but based on conversations with caregivers who told me the things they wish they had known:

Where to find support and education programs in your local area. It's vital to have a good support network for advice, encouragement, and knowledge. You need to know what to expect and how to prepare to meet the challenges that lie ahead.

Where to find early-stage social engagement programs. These are programs that help people in the early stages of the disease stay connected and active. A diagnosis does not mean a person disengages from life and is destined to be confined to a lounge chair in the living room or in a facility. Adult day health care centers that specialize in serving people with dementia are cropping up in many cities. Programs that include cognitive rehabilitation therapy should also be considered. These offer a wide range of therapies provided by trained professionals that help people relearn skills they have lost through traumatic brain injuries or the cognitive decline seen in dementia. New research shows that cognitive rehab may be able to teach people to compensate for memory and thinking shortfalls that happen in the early stages. Remember that what you do during those early stages may have a significant impact on how fast the disease progresses. And the diagnosis does not mean you stop learning new things. Some people thrive for a long time and can even continue to live independently with the right support in place.

Where to find clinical trials matching your needs. These studies will help you become part of important research, but they may also slow the progression of the disease. There is no guarantee with any clinical study that you'll find a helpful treatment, let alone a cure, but rarely will there be downsides to participating.

How to keep a home safe. People in the early stages often lead an independent life, but there will be preparations and hard choices to make, such as no more driving and walking outside alone. At some point, a person with progressing dementia will need help with daily tasks such as managing money and paying bills, shopping and cooking, general housework, and personal tasks like grooming, dressing, bathing, toileting, and taking medication. Eventually, the home, no matter how many safety features have been set up, may not be the ideal place to stay. Where will you go? AARP's book *Wise Moves* can help you choose among the options.

How to make a legal plan. This includes taking inventory of the family's legal documents—wills and trusts. If none are in place, a family or estate attorney can help draft and execute these important documents, which include things like durable power of attorney (designating who can make financial and other decisions when someone is no longer able) and durable power of attorney for health care (designating who can make health care decisions when someone is no longer able). They are valid even after the person is no longer able to make decisions. These documents tend to be lengthy and detailed, and they specify some of the most practical but difficult decisions that will eventually be faced, such as care facilities, types of treatment, end-of-life care decisions (e.g., do you want feeding tubes?), and DNR (do not resuscitate) orders. These are important decisions to make because without instructions in place, expensive medical interventions are often routinely performed even if they are futile in extending life. One young woman told me about her own mother, "Life itself quickly became a transactional physical and financial death spiral devoid of emotion." Think of it this way: You've worked hard your entire life to build a little wealth and have something to leave to others. But if you don't plan, all of those assets can vanish because of the costs incurred during the last part of your life.

How to make a financial plan. This part of the process can be daunting and will have some crossover with legal planning. You'll want to organize assets, debts, insurance policies, and existing benefits like Medicare, retirement, and Social Security. The Financial and Legal Document Worksheet on the Alzheimer's Association's website can help you in your inventory. As part of this exercise, you'll also want to identify the cost of care going forward—from basics like ongoing medical treatment and prescription drugs to adult day services, in-home care services, full-time residential care services, and the prospect of moving to a facility that specializes in end-stage Alzheimer's. There will be many options to explore. If this part of the planning feels overwhelming and uncomfortable or you're dealing with a complex family estate, it helps to bring in a financial advisor who is qualified (and licensed and certified) to be your guide. Be sure to choose this person carefully—preferably someone who has gone down this path with many families and is knowledgeable about eldercare and long-term care planning. The Alzheimer's Association provides leads and links to directories on its website for finding this important person if you don't already have someone on board or in mind. Another good resource is AARP's *Checklist for My Family: A Guide to My History, Financial Plans, and Final Wishes.*

How to build a care team. No one can walk this road alone. In addition to family, your friends, neighbors, and health care professionals are all part of your team. Volunteers in your community can also be part of this team. The sooner you identify and develop your care team upon diagnosis, the better. These conversations can be challenging, especially when you are not quite ready to share your diagnosis widely. Over and over again, though, experts tell me that having these people in your inner circle early on will allow you to live as full of a life as possible and for as long as possible. Again, choose these people wisely!

Q: I was told to get an advance directive. What is that?

A: Advance directives are legal documents that allow people to document their wishes regarding treatment and care, including end-of-life preferences. They include a living will that dictates how you'd want your end-of-life care to go and who will be in charge, or your durable power of attorney for health care. Fewer than 30 percent of American adults have signed advance directives stating their health care wishes. Here is why this is so important. Forgoing these documents can be financially ruinous to families—triggering unexpected medical bills and bankruptcies to loved ones left to deal with the financial aftermath. When a person does not have an advance directive, the costs can skyrocket. According to the Agency for Healthcare Research and Quality, one-quarter of all Medicare spending annually—$139 billion—each year goes toward care for just 5 percent of beneficiaries at the end of life.[5] Put another way, 25 percent of Medicare's annual spending is used by 5 percent of patients during the last twelve months of their lives. Advance directives also can help to avoid unwanted, ineffective medical interventions that can cause so much anguish for loved ones. The data is alarming: Spouses are *twice as likely to prematurely die* after their partner's death if end-of-life care is not planned.[6] How so? I strongly believe it's the stress. When we think about the costs of dealing with dementia, we forget about the costs that don't come with dollar signs: Nearly 60 percent of those caring for family members with Alzheimer's or other dementias report "high" or "very high" emotional stress.

Q: I don't have a large estate so I don't need a trust, do I? Aren't those for rich people?

A: Anyone who owns property and assets—from a house to bank accounts—should have a will or trust in place; these are not documents only for the super wealthy. If you have substantial assets, ideally you set up what's called a living trust while you are still alive that pools all of your assets into one entity, the trust, so your family can avoid the long and often costly probate process the courts use to distribute your assets after your death. In setting up a trust, you make instructions about how you'd like your assets to be handled once you are no longer able to manage your affairs, and you name a trustee to follow those instructions, with a backup successor trustee as well. Living trusts and wills often get drafted together as a package. The costs of not having these documents in place before you die depend on the state in which you live. But in some states, dying without a will (intestate) and trust can be devastating to your beneficiaries—and your legacy. A large estate can be wiped out by the probate process, lawyers, and bickering family members who don't agree on how to divide your assets.

KEEP TALKING

To be clear, everyone should complete these documents, not just those worried about dementia. When Nancy's father passed without a will or trust in place, she and her siblings fought to figure out how to take care of their mother, who was in the middle stages of Alzheimer's and could not live independently or make decisions for herself (what's technically called lacking "legal capacity"). No plans had been set and no one could agree on the best steps forward for their mother's sake. One sibling thought their mom should be placed in an assisted living facility that catered to people with dementia. Another strongly believed she should stay in the home at whatever cost and hire around-the-clock care when necessary. The third daughter had mixed feelings about all the options and could not take sides. The debate grew bitter and prolonged while their mother suffered. Eventually one of the siblings filed a petition with the court asking for a conservator to come in and act as a leader. This is not common, but when families cannot agree about how to handle an individual's legal, financial, or health care decisions, the courts can get involved. In some states, a conservator is called a guardian.

Conservatorships are not typically ideal solutions. They involve court proceedings, additional costs, and lawyers, and you or your family members may lose control, not even getting to choose who becomes the conservator and how things get decided in the future. Every state has different laws that govern this area of family law, but it tends to be riddled with problems and a general lack of oversight that may allow for unscrupulous behavior among conservators. According to fiduciary watchdogs and family lawyers who attend conservatorship proceedings, people who are cognitively incapacitated and have warring family members are extremely vulnerable. In many cases throughout the country, large estates have been drained by this system and elders with dementia are preyed on and financially abused. Indeed, a conservator is supposed to "conserve" an estate and protect the individual; the same goes for guardians, as they are named in some states. But conservators and guardians can hold so much power that they not only make all the decisions when it comes to an individual's health care and well-being, but they can decide the fate of assets, property, and even where the person lives—without the family's input or wishes. Conservators and guardians often are granted trustee status, too, further extending their power. Once a conservator or guardian is attached to an estate, it can be incredibly difficult to end or object to the conservatorship or guardianship without arduous and expensive court hearings. These proceedings in general are emotionally exhausting and can be grueling for family members who are already under the stress of having to deal with each other's quarrels and a loved one's dementia.

The best way to avoid a court-appointed conservator or guardian is to have open communication with family members early and often. Make this a priority; do it as soon as you finish reading this book. Have your will or trust prepared. I realize that communication can be tough in some families, and a diagnosis of dementia complicates matters, but it's essential. Plan a family meeting and bring in a trusted family friend if that helps for extra support. This might require multiple family meetings and that's okay. If everyone cannot attend the meeting in person, use tools like voice-over IP or Skype to make sure everyone can be included.

THE INVISIBLE SECOND PATIENT

Here is a statistic that I initially found hard to believe: Caregivers of spouses with dementia are up to six times more likely to develop dementia than people in the general population.[7] In fact, anyone who helps care for a loved one with dementia has a higher risk of developing the ailment. These people are called the "invisible second patients." It seems ironic and cruel, but it makes sense if you consider the dynamic. The spouse caregiver has been, on average, married for thirty years, and there is now a significant upheaval in the couple's shared life. On top of that are increased stress, loneliness, depression, and inactivity. Their devotion to such care often means a trade for a lower quality of life. And as I heard many times, the emotional effects of witnessing the disease progress despite your care and support creates a sense of profound helplessness.

We hear about toxic stress in the media daily and its biological effects in the body—from the destructive slow boil of chronic inflammation to increased stress hormones like cortisol that inflict biological harm over time. I have reported on the ills of toxic stress in America, mostly among communities where economic divides run deep due to income inequalities and a lack of general optimism about the future. This high-anxiety state can result in drug dependence, suicide, and a heightened risk of dying from illnesses like cardiovascular disease or stroke. But we don't think about the same toxic stress experienced by caregivers who often suffer similar emotional and physical consequences. The biological reasons for their elevated risk of developing dementia are partly the same: Chronic inflammation ravages the body and reaches the brain. In fact, caregivers are at increased risk not only for dementia but also for any ailment tied to chronic inflammation, which is every degenerative disease we know of today, from heart disease to cancer.

When we think of dementia, we typically envision some variation of "forgetfulness disorder" in our heads. We don't often consider the other symptoms that often come with dementia and can be excruciatingly hard to manage, particularly for the caregivers. These include anger, agitation, mood changes, hallucinations, apathy, sleep disturbances, incontinence, and wandering. These challenging dementia-related symptoms, in fact, are some of the leading reasons people are placed in assisted living facilities or nursing homes. It becomes too hard and stressful to care for them. Those of us who are parents can vividly remember the sleepless days of having an infant in the house who has yet to maintain a regular sleep schedule. But we know those days are numbered, and soon enough our babies will be toddlers with stable sleeping patterns. Now imagine what it's like to be responsible for an adult who no longer maintains a constant, reliable sleeping schedule. She sleeps randomly throughout the day and night, sometimes waking every couple of hours when everyone else in the house is asleep. Add to that troubles with eating, using the bathroom, and walking (incontinence is another leading reason people are placed in nursing homes). And her personality can morph with the disease. Someone who was mean can become, after developing dementia, gentle and sweet. Someone who was once loving, easygoing, and fun to be around can become increasingly cantankerous, combative, lacking in social graces, and likely to have unpredictable outbursts. Caregivers can feel like they are walking on eggshells and don't know what they will encounter when they walk into the room to see their loved one. These behaviors can get worse with time, and with patients who wander at night or act out hallucinations, the situation can quickly become intolerable and untenable. Unfortunately, there's no way to predict who will experience these challenging behaviors and symptoms, which can change depending on the stage of the disease process, the situation, and what area of the person's brain is most affected by the illness.

In the early stages, when a person's cognition is only mildly impaired but the person has awareness of what is happening, anxiety, anger, aggression, and mild depression can occur. As many as 20 percent of individuals with Alzheimer's will experience increased confusion, anxiety, restlessness, and agitation beginning late in the day. This is called sundowning or sundowner's syndrome. Later in the disease, once the dementia renders someone less aware of his mood shifts, paranoia, delusions, and hallucinations can set in. There are no effective treatments for these kinds of symptoms, and antipsychotic drugs are sometimes associated with an increased risk of death in people with dementia. While there is always great interest in developing effective treatments for the disease itself, there is also hope that research will generate better strategies to help combat these most devastating symptoms with safer drugs or even drug-free approaches. For example, promising research is underway to look at using the effect that light has on the body's sleep-wake cycles. The thinking is that improving the sleep patterns of dementia patients could significantly improve their mood and behavior.

Q: My mother is delusional and hallucinating. She accuses me of all sorts of things, from stealing to killing people. Is this normal? What do I do?

A: In the middle to late stages of Alzheimer's, delusions and hallucinations can occur. These are not the same thing. Delusions are firmly held beliefs that are not real, such as a suspicious delusion that someone is stealing possessions. This is sometimes called a paranoia. Hallucinations are false perceptions of events or objects that are sensory in nature. This is when an Alzheimer's patient sees, smells, tastes, hears, or feels something that is not there. A caregiver who encounters either delusional thinking or hallucinations in a person with Alzheimer's should document the specific behavior as much as possible to share with the doctor. Witnessing these experiences can be deeply troubling, and sometimes the patient can act out in ways that hint at self-harm or caregiver harm. There may be some treatment options to consider at this juncture, depending on the specific symptoms and the stage of dementia.

DON'T FORGET YOURSELF: A NOTE FOR CAREGIVERS

Taking care of a loved one with dementia needs to be a team effort with family members and friends. But for the person who takes on the role of primary caregiver (and there always is one), it's crucially important that he or she prioritize self-care in addition to the care of the patient. This means staying on top of your own diet and exercise routines, engaging in activities that boost your well-being, spending time with friends and family, and taking breaks from your caregiving duties—breaks throughout the day (even just five minutes) and longer breaks throughout the weeks and months with certain days and weekends off. The program I outlined in part 2 is designed for all of us—whether we are already caring for someone, awaiting a diagnosis in a loved one, or headed toward serious cognitive decline ourselves. Put yourself on your to-do list.

If you're a caregiver who also maintains a full- or part-time job, be extra cautious about your time, energy, emotions, and personal needs. You are at a high risk for burnout, but not for reasons you may think. Caregiver burnout is caused less by the rigorous responsibilities of the jobs themselves and more by the fact you tend to neglect your own emotional, physical, and spiritual health. To repeat, put yourself on your to-do list. Listen to any symptoms you develop and pay attention to them. While things may seem okay at the start of your dual duties, you can't know for sure how long you will have to wear both hats. This can be a long and punishing road, leading to further neglect of self-care. By the time most caregivers are burned out, they are sick themselves.

Don't be ashamed to recruit help for yourself and your loved one. Again, talk to your siblings and anyone else in your circle who can assist you. I've witnessed too many people wait far too long to ask for help, and they wind up with serious health problems of their own that can be just as devastating—or more so—as their partner's dementia. In one tragic instance, the wife died of a fatal heart attack while caring for her husband as he deteriorated with a difficult form of dementia. She was trying to do everything on her own, not wanting to "burden" or "bother" anyone else. I wonder if she'd still be alive today had she gotten more help and been able to take better care of herself.

Family caregivers are motivated to provide care for any number of reasons—from a sense of love or reciprocity to a sense of guilt or duty. It may be helpful to identify your specific motivation so it can serve as a reminder when things are particularly challenging. There are social pressures and cultural norms to meet. In rare instances, greed can be a motivator, but that is not common. Many people tell me that caring for a loved one at this time in life can be incredibly fulfilling on a spiritual level. It comes as no surprise, however, that caregivers who are motivated by the negative forces—duty, guilt, or social pressure—are more likely to resent their role and suffer greater psychological distress than those who are motivated by positive incentives. And those who identify more with beneficial components of their role experience less burden, better health and relationships, a more gratifying experience, and greater social support.

One of the hardest things a caregiver faces, at least in the beginning, is denial. And that's totally normal. It's not easy learning that a parent, partner, or other family member has a disease as scary and fateful as Alzheimer's. We don't get this kind of preparation in formal schooling. Even in medical school, I wasn't taught the fundamentals of dealing with the psychological aspect to hearing about a family member's grim diagnosis. I've learned a lot through my years as a practicing physician and talking to families who are grappling with a difficult prognosis, and I have dealt with it personally with my parents and a grandparent. It is always hard. The diagnosis can seem unbelievable, impossible to accept. Your life is probably already overscheduled with responsibilities, and then you add something that requires almost another full-time commitment. It's no surprise that denial in the short term can actually be a healthy coping mechanism, as it provides time to get used to the new reality and let circumstances sink in. But you can't stay in denial forever, especially when there are decisions and plans to make. If you cannot come to terms with the diagnosis, talk to someone and seek professional help from a therapist. Diagnoses like these can be incredibly disruptive to self-confidence as well, and a therapist can help you reframe your thoughts a certain way to regain the confidence that you'll need to move forward.

Guilt is another emotion that many experience at the beginning, and it rides right along with the denial. You wonder why you didn't see the signs earlier and question why you avoided seeing them. Could your loved one have been better off if you got her diagnosed sooner and into treatment? These emotions—denial and guilt—are common. But again, it's important to stay attuned to your own emotions and mental and physical exhaustion, and equip yourself with knowledge and resources. Connect with other caregivers who are in a similar situation.

I can't emphasize this enough: Build your own support network, ask for and accept help, and continually plan for the future, adjusting plans as needed and being okay with uncertainty. Alzheimer's disease is erratic, unpredictable, and an instigator of many mixed emotions—anxiety, fear, sadness, depression, anger, frustration, and grief. Try to be mindful of what you're feeling and respond to your own needs. Keep in mind that the disease can vary tremendously from person to person and progress differently across the stages. So don't beat yourself up when you compare symptoms with others and find that you "have it worse" than another caregiver. Embrace and accept the fact you've taken on one of the most challenging roles anyone can have in life. Both AARP and the Alzheimer's Association provide a wealth of resources online about care options, caregiving across the stages, support, and financial and legal planning for caregivers. The sites also include a bounty of strategies for handling complicated situations for which none of us has likely been formally trained. Indeed, there are ideal ways to respond when faced with someone behaving out of character or extremely unpredictably. It can be hard to know how to regularly monitor that person's comfort and anticipate needs when things change so quickly. It also can be excruciatingly tough to know how to handle unusual situations. For example, what should you do when the person seems stuck on repeating a word, activity, or sentence over and over again? Repetition is common in the disease's later stages. The person is searching for familiarity and comfort as the brain continues its malicious march forward in decline. One of the ways to respond, in addition to being calm and patient, is to engage the person in an activity to break the pattern of repetition. The Alzheimer's Association's website has online support community and message boards (ALZConnected) where people can share their strategies. It helps to share your experience with others. This is a team effort in terms of family and a group effort in global terms.

The goal of the caregiver is ultimately to help the person with dementia live well. It is a job high in demand and low in gratitude or compensation. I don't know if there is such a thing as true balance, but I will say that the role is very much a balancing act.

You may find that at some point, you can no longer be the primary caregiver. Be open to the idea of changing the setting for the patient and give yourself permission to surrender the sole responsibility. Again, there are lots of options out there for places that offer high-quality care led by professionals specially trained to treat people with dementia and who do so with respect and dignity. Don't get to a place where you feel trapped and resentful. All that's asked of you is to do what you can when you can. It may help to keep a journal for your personal thoughts and notes. Keep track of your experience. Document the journey.

CONCLUSION

The Bright Future

*The future enters into us, in order to transform
itself in us, long before it happens.*

RAINER MARIA RILKE

I promised that I'd end this book on an optimistic high note. In the time that has passed since I wrote these words and you're reading this sentence, thousands of headlines will have been published with the word *Alzheimer's*. There is no shortage of enthusiasm and drive to find better treatments, perhaps a cure. In 2019, the possibility of a vaccine once again burst onto the scene after scientists at the University of New Mexico reported on their experiments of inoculating mice with a virus-like particle designed to target tau protein. The mice developed antibodies that removed the abnormal tau protein from the part of the brain associated with learning and memory. Will it work in humans and have antidementia effects? It remains to be proven.

Another team of scientists is hard at work with so-called endobody vaccines—vaccines that prime the immune system to deal with malfunctioning internal parts of the body that it would otherwise ignore. These work differently from the typical vaccine that prepares the body's immune system to fight off diseases that come from the external world, such as flu or measles, which are caused by bacteria or viruses entering the blood. The endobody vaccines essentially provoke an antibody response in the body that clears the tangled beta-amyloid plaques without triggering damaging inflammation. Clinical trials are underway to see if this vaccine will have an impact on cognition and memory, but it will probably take years before results are in. And yet another group, this one out of Yale University, has suggested that a "drinkable cocktail of designer molecules" can restore memories in mice engineered to have a condition similar to Alzheimer's disease. Science fiction or a potential therapy? Future research will answer this question. Future research will also help us put an end to a range of illnesses linked to the brain, from mental disorders such as depression, anxiety, bipolar disorder, and schizophrenia to neurodegenerative maladies such as Parkinson's and amyotrophic lateral sclerosis, or Lou Gehrig's disease. Although each of these illnesses is unique, my hunch is that breakthroughs in treating or curing one ailment will inform other areas of brain science. What we learn from studying depression, for example, may help us better understand Alzheimer's disease. There are lots of surprising intersections in medicine. We just need to find them.

I'm excited for what the future will bring us in our understanding and treatment of diseases as complex as Alzheimer's and other forms of dementia. Even that word, *dementia*, may one day be forgotten. With new therapies on the horizon, I don't think it will be fair to label anyone with "dementia" if they can go on to live a life with an ailment that is kept at bay. Our entire vocabulary and the narrative around degenerative brain diseases will change with promising new preventive solutions and treatments for symptoms. Preventing and treating brain ailments will not be reduced to a single action, but will entail a multipronged approach. The solutions will probably encompass an array of things, from modifiable lifestyle strategies and daily habits to medications and gene therapies.

I hope that I've given you a lot to think about and do in pursuit of a vibrant brain. My teenaged kids will probably be among the first of many generations to come who will push the limits of human longevity—living long and sharply into their nineties and beyond. With the dawn of personalized medicine upon us and the explosion of new drugs and therapies that can revolutionize and democratize medicine, we're on the precipice of a new era in our evolution as a species. The pace of change will only grow faster. Imagine a retinal scan through your smartphone or iPad that tells you which mix of molecules or biologics will clear your brain of suspicious proteins, restore synapses, and heighten cognition. Or picture a drone delivering the right therapy to the right person at the right time that will enhance the brain's processing speed without side effects. We will soon be able to peer into our brains and see where a problem is developing while we are armed with small molecules or natural plants to help address that problem. I am convinced that we have created many of the problems that plague us, and this presents an opportunity. There will always be a place for good old-fashioned habits like eating more vegetables and working out regularly. But those time-tested habits coupled with what's in store for us tomorrow will ultimately make for the best life—one that we will want to remember and will be able to remember. Keep sharp.

Acknowledgments

The scientists who wake up every morning with the belief that diseases are not preordained, that memory loss need not accompany aging, and that everyone can make their brains better inspired me to write this book. Over nearly two decades, I would speak to these scientists at the big brain meetings, in their laboratories and in their homes. They would share their scientific findings but also let me in on the deeply personal reasons they had chosen to study the brain in the first place. They convinced me not only that we would one day make diseases like dementia a thing of the past, but that even a healthy brain could be improved and made more resilient. Thank you for your candor and your willingness to help take some of the most remarkable new knowledge about the brain and make it relevant for anyone, anywhere.

Priscilla Painton, executive editor is your title but it does not nearly begin to describe the role you have played in this book. From the beginning, your vision was clear and your collaboration far exceeded my expectations. Your remarks and notes were always spot-on, and always added great value. You have an ability to see around corners and anticipate the direction of the book. I have been fortunate to have such a dedicated and professional team working on *Keep Sharp*, and we have also become a family along the way. Richard Rhorer, Julia Prosser, Elizabeth Gay, Elise Ringo, Yvette Grant, Carly Loman, Jackie Seow, Lisa Erwin,

Marie Florio, Hana Park, and finally Megan Hogan, who now holds a land speed record on returning emails, and always with a smile. Thanks to all of you.

Jonathan Karp, you are the definition of a gentleman and a scholar. I knew after the first meeting in your office, when we discussed everything from stem cells to Springsteen, that I was dealing with someone truly engaged with the world. Thanks for believing in me and this book.

Bob Barnett is a world-famous lawyer. He has represented presidents and the pope. Yet you would never know it. He is so incredibly humble and hardworking. One of the best days in my life was the day Bob Barnett agreed to help me with my career. His guidance has been remarkably prescient and insightful.

The collaboration I have had with my partner and friend, Kristin Loberg, has been truly speical. We should all be lucky enough to have a real mind meld with someone like Kristin, who immediately understood what I was trying to convey and always helped me get there. She is the very best at what she does, and quite simply, this book would not have been possible without her.

Notes

The notes that follow offer a partial list of scientific papers and other references that you might find helpful if you want to learn more about some of the ideas and concepts expressed in this book. I have cited the studies mentioned in the book. If I had my wish, I'd cite every paper I've read on this subject, but the list would come to thousands of entries. At the least, these materials can open doors for further research and inquiry.

INTRODUCTION

1 M. A. Rivka Green, Bruce Lanphear, and Richard Hornung et al., "Association between Maternal Fluoride Exposure during Pregnancy and IQ Scores in Offspring in Canada," *JAMA Pediatrics*, August 19, 2019. doi:10.1001/jamapediatrics.2019.1729. [Epub ahead of print.]

2 Matthew J. Burke, M. Fralick, N. Nejatbakhsh, et al., "In Search of Evidence-Based Treatment for Concussion: Characteristics of Current Clinical Trials," *Brain Injury* 29, no. 3 (November 2015): 300–305.

3 R. Brookmeyer, N. Abdalla, C. H. Kawas, and M. M. Corrada, "Forecasting the Prevalence of Preclinical and Clinical Alzheimer's Disease in the United States," *Alzheimer's & Dementia* 14, no. 2 (February 2018): 121–129.

4 For updated numbers and figures on the prevalence of Alzheimer's disease,

among other brain ailments, see the Alzheimer's Association (www.alz.org) or the Centers for Disease Control and Prevention (www.cdc.gov).

5 Jeffrey L. Cummings, Travis Morstorf, and Kate Zhong, "Alzheimer's Disease Drug-Development Pipeline: Few Candidates, Frequent Failures," *Alzheimer's Research and Therapy* 6, no. 4 (July 2014): 37.

6 Nao J. Gamo, Michelle R. Briknow, Danielle Sullivan, et al., "Valley of Death: A Proposal to Build a 'Translational Bridge' for the Next Generation," *Neuroscience Research* 115 (February 2017): 1–4.

7 J. G. Ruby, K. M. Wright, K. A. Rand, et al., "Estimates of the Heritability of Human Longevity Are Substantially Inflated due to Assortative Mating," *Genetics* 210, no. 3 (November 2018): 1109–1124.

PART 1: THE BRAIN

1 We often hear that there are as many—if not more—neurons in the human brain as stars in the Milky Way. This is a very generalized analogy used to convey a sense of enormity and scale, though technically we don't really know exact numbers for either neurons or stars in our galaxy. The most recent estimates approximate 86 billion neurons in the human brain and 200 to 400 billion stars in the Milky Way. So perhaps the stars outnumber our brain cells. But again, the analogy is not meant to be taken literally and methods to arrive at these numbers are not without their flaws. For an interesting explanation of this conundrum, see Bradley Voytek's article for *Nature*: "Are There Really as Many Neurons in the Human Brain as Stars in the Milky Way?" May 20, 2013.

2 This quote is attributed to James D. Watson and is written in the foreword to Sandra Ackerman's *Discovering the Brain* (Washington, DC: National Academies Press, 1992).

CHAPTER 1: WHAT MAKES YOU *YOU*

1 Numbers often cited for the average surface area of a human cortex reflect a range from 1.5 square feet to over 2 square feet. For a review paper on this subject, see Michel A. Hofman's "Evolution of the Human Brain: When Bigger Is Better," *Frontiers in Neuroanatomy* 8 (March 2014): 15.

2 To date, there has never been a peer-reviewed journal to corroborate the 100-billion neuron fact. This is an estimation based on informal interpolations from various measurements. Interestingly, Suzana Herculano-Houzel and her col-

leagues published a paper in 2009 that showed a calculation of 86 billion using a novel way to count them. See: "Equal Numbers of Neuronal and Nonneuronal Cells Make the Human Brain an Isometrically Scaled-up Primate Brain," *Journal of Comparative Neurology* 513, no. 5 (April 2009): 532–41. Also check out her TED talk on the subject: www.ted.com/speakers/suzana_herculano_houzel.

3 John M. Harlow, "Recovery from the Passage of an Iron Bar through the Head," *Publications of the Massachusetts Medical Society* 2, no. 3 (1868): 327–47. Reprinted by David Clapp & Son (1869).

4 To access a library of data and information about the brain, see www.BrainFacts.org.

CHAPTER 2: COGNITIVE DECLINE–REDEFINED

1 Michelle Cortez, "Merck Stops Alzheimer's Study After 'No Chance' of Benefit," *Bloomberg Business*, February 14, 2017.

2 G. S. Bloom, "Amyloid-β and Tau: The Trigger and Bullet in Alzheimer Disease Pathogenesis," *JAMA Neurology* 71, no. 4 (April 2014): 505–508.

3 To follow Dr. Stern's research, go to his academic website: www.bu.edu/cte/about/leadership/robert-a-stern-ph-d.

4 Lulit Price, Christy Wilson, and Gerald Grant, "Blood–Brain Barrier Pathophysiology following Traumatic Brain Injury," in *Translational Research in Traumatic Brain Injury* (Boca Raton, FL: CRC Press/Taylor and Francis Group, 2016), 85–96.

5 A. Montagne, S. R. Barnes, M. D. Sweeney, and M. R. Halliday, "Blood-Brain Barrier Breakdown in the Aging Human Hippocampus," *Neuron* 85, no. 2 (January 2015): 296–302.

6 Maria Aguilar, Taft Bhuket, Sharon Torres et al., "Prevalence of the Metabolic Syndrome in the United States, 2003–2012," *JAMA* 313, no. 19 (May 2015): 1973.

7 Owen Dyer, "Is Alzheimer's Really Just Type III Diabetes?" *National Review of Medicine* 2, no. 21 (December 2005). www.nationalreviewofmedicine.com/issue/2005/12_15/2_advances_medicine01_21.html.

8 H. J. Lee, H. I. Seo, H. Y. Cha, et al., "Diabetes and Alzheimer's Disease: Mechanisms and Nutritional Aspects," *Clinical Nutrition Research* 7, no. 4 (October 2018): 229–240.

9 Fanfan Zheng, Li Yan, Zhenchun Yang, et al., "HbA1c, Diabetes and Cognitive Decline: The English Longitudinal Study of Ageing," *Diabetologia* 61, no. 4 (April 2018): 839–848.

10 N. Zhao, C. C. Liu, A. J. Van Ingelgom, and Y. A. Martens, "Apolipoprotein E4 Impairs Neuronal Insulin Signaling by Trapping Insulin Receptor in the Endosomes," *Neuron* 96, no. 1 (September 2017): 115–129.e5.

11 R. A. Whitmer, E. P. Gunderson, E. Barrett-Conner et al., "Obesity in Middle Age and Future Risk of Dementia: A 27 Year Longitudinal Population Based Study," *British Medical Journal* 330, no. 7504 (June 2005): 1360.

12 C. C. John, H. Carabin, S. M. Montano et al., "Global Research Priorities for Infections That Affect the Nervous System," *Nature* 527, no. 7578 (November 2015): S178–186.

13 Bret Stetka, "Infectious Theory of Alzheimer's Disease Draws Fresh Interest," Shots: Health News from NPR September 9, 2018. See www.npr.org/sections/health-shots/2018/09/09/645629133/infectious-theory-of-alzheimers-disease-draws-fresh-interest.

14 W. A. Eimer, D. K. Vijaya Kumar, N. K. Navalpur Shanmugam et al., "Alzheimer's Disease-Associated β-Amyloid Is Rapidly Seeded by Herpesviridae to Protect against Brain Infection," *Neuron* 99, no. 1 (July 2018): 56–63.

15 K. A. Walker, R. F. Gottesman, A. Wu et al., "Systemic Inflammation during Midlife and Cognitive Change over 20 Years: The ARIC Study," *Neurology* 92, no. 11 (March 2019): e1256–e1267.

16 C. Zhang, Y. Wang, D. Wang et al., "NSAID Exposure and Risk of Alzheimer's Disease: An Updated Meta-Analysis from Cohort Studies," *Frontiers in Aging Neuroscience* 10 (March 2018): 83.

17 www.alz.org.

18 M. Boldrini, C. A. Fulmore, A. N. Tartt et al., "Human Hippocampal Neurogenesis Persists throughout Aging," *Cell Stem Cell* 22, no. 4 (April 2018): 589–599.

19 These figures are from the Alzheimer's Association and are based on long-term studies.

20 See the Alzheimer's Association's annual "Disease Facts and Figures" at www.alz.org.

21 Ibid.

22 Ibid.

23 Mary A. Fischer, "6 Types of Normal Memory Lapses and Why You Needn't

Worry About Them," AARP, stayingsharp.aarp.org/about/brain-health/normal -memory/.

24 Harry Lorayne and Jerry Lucas, *The Memory Book: The Classic Guide to Improving Your Memory at Work, at School, and Play*, reissue ed. (New York: Ballantine Books, 1996).

25 For a review, see Cheryl Grady, "Trends in Neurocognitive Aging," *Nature Reviews Neuroscience* 13, no. 7 (June 2012): 491–505.

26 Majid Fotuhi, "Changing Perspectives Regarding Late-Life Dementia," *Nature Reviews Neurology* 5 (2009): 649–658.

CHAPTER 3: 12 DESTRUCTIVE MYTHS AND THE 5 PILLARS THAT WILL BUILD YOU

1 L. Rena and Meharvan Singh, "Sex Differences in Cognitive Impairment and Alzheimer's Disease," *Frontiers in Neuroendocrinology* 35, no. 3 (August 2014): 385–403.

2 M. Colucci, S. Cammarata, A. Assini et al., "The Number of Pregnancies Is a Risk Factor for Alzheimer's Disease," *European Journal of Neurology* 113, no. 12 (December 2006): 1374–1377.

3 E. E. Sundermann, A. Bigon, L. H. Rubin et al., "Does the Female Advantage in Verbal Memory Contribute to Underestimating Alzheimer's Disease Pathology in Women versus Men?" *Journal of Alzheimer's Disease* 56, no. 3 (February 2017): 947–957.

4 Keith A. Wesnes, Helen Brooker, Clive Ballard et al., "An Online Investigation of the Relationship between the Frequency of Word Puzzle Use and Cognitive Function in a Large Sample of Older Adults," *International Journal of Geriatric Psychiatry* 34, no. 7 (2018): 921–931. Helen Brooker, Keith A. Wesnes, Clive Ballard et al., "The Relationship between the Frequency of Number Puzzle Use and Baseline Cognitive Function in a Large Online Sample of Adults Aged 50 and Over," *International Journal of Geriatric Psychiatry* 34, no. 7 (July 2019): 932–940.

5 P. S. Eriksson, E. Perfilieva, T. Björk-Eriksson T. et al., "Neurogenesis in the Adult Human Hippocampus," *Nature Medicine* 4, no. 11 (November 1998): 1313–1317.

6 Sharon Begley, *Train Your Mind, Change Your Brain: How a New Science Reveals Our Extraordinary Potential to Transform Ourselves* (New York: Ballantine, 2007).

7 See www.johnratey.com.

8 Michael Merzenich, *Soft-Wired: How the New Science of Brain Plasticity Can Change Your Life*, 2nd ed. (San Francisco: Parnassus Publishing, 2013).

9 This quote was written by Michael Merzenich and a colleague in 1996, though it never appeared in a peer-reviewed journal. It's best memorialized in Sharon Begley's *Train Your Mind, Change Your Brain: How a New Science Reveals Our Extraordinary Potential to Transform Ourselves* (New York: Ballantine, 2007), 159.

10 Matthew J. Huentelman, Ignazio S. Piras, Ashley L. Siniard et al., "Associations of MAP2K3 Gene Variants with Superior Memory in SuperAgers," *Frontiers in Aging Neuroscience* 10 (May 2018): 155.

11 D. C. Park, J. Lodi-Smith, L. Drew et al., "The Impact of Sustained Engagement on Cognitive Function in Older Adults: The Synapse Project," *Psychological Science* 25, no. 1 (January 2014): 103–112.

12 See the work of Earl Keith Miller and the Miller Lab: http://millerlab.mit.edu.

13 T. Molesworth, L. K. Sheu, S. Cohen et al., "Social Network Diversity and White Matter Microstructural Integrity in Humans," *Social Cognitive and Affective Neuroscience* 10, no. 9 (September 2015): 1169–1176.

PART 2: THE BRAIN TRUST

1 The Alzheimer's Association, www.alz.org.

2 Ibid.

CHAPTER 4: THE MIRACLE OF MOVEMENT

1 R. C. Petersen, O. Lopez, M. J. Armstrong et al., "Practice Guideline Update Summary: Mild Cognitive Impairment: Report of the Guideline Development, Dissemination, and Implementation Subcommittee of the American Academy of Neurology," *Neurology* 90, no. 3 (January 2018): 126–135.

2 D. E. Barnes and K. Yaffe, "The Projected Effect of Risk Factor Reduction on Alzheimer's Disease Prevalence," *Lancet Neurology* 10, no. 9 (September 2011): 819–828.

3 P. F. Saint-Maurice, D. Coughlan, S. P. Kelly et al., "Association of Leisure-Time Physical Activity across the Adult Life Course with All-Cause and Cause-Specific Mortality," *JAMA Network Open* 2, no. 3 (March 2019): e190355.

4 S. Beddhu, G. Wei, R. L. Marcus et al., "Light-Intensity Physical Activities and Mortality in the United States General Population and CKD Subpopulation,"

Clinical Journal of the American Society of Nephrology 10, no. 7 (July 2015): 1145–1153.

5　D. M. Bramble and D. E. Lieberman, "Endurance Running and the Evolution of Homo," *Nature* 432, no. 7015 (November 2004): 345–352.

6　Daniel Lieberman, *The Story of the Human Body: Evolution, Health, and Disease* (New York: Pantheon, 2013).

7　D. E. Lieberman, "Is Exercise Really Medicine? An Evolutionary Perspective," *Current Sports Medicine Reports* 15, no. 4 (July–August 2015): 313–319.

8　Lieberman, *The Story of the Human Body*: p. 6.

9　C. M. Tipton, "The History of 'Exercise Is Medicine' in Ancient Civilizations," *Advances in Physiology Education* 38, no. 2 (June 2014): 109–117.

10　Susruta Susruta and Kunja Lal Bhishagratna, *An English Translation of the Sushruta Samhita, Based on Original Sanskrit Text*; Vol. 1–3 (Franklin Classics, 2018).

11　For a well-cited review of all the benefits of exercise, go to the National Institutes of Health's U.S. National Library of Medicine online and access its Medline Plus "Benefits of Exercise" topic at medlineplus.gov/benefitsofexercise.html.

12　K. Segaert, S. J. E. Lucas, C. V. Burley et al., "Higher Physical Fitness Levels Are Associated with Less Language Decline in Healthy Ageing," *Scientific Reports* 8, no. 1 (April 2018): 6715.

13　S. Chetty, A. R. Friedman, K. Taravosh-Lahn et al., "Stress and Glucocorticoids Promote Oligodendrogenesis in the Adult Hippocampus," *Molecular Psychiatry* 19, no. 12 (December 2014): 1275–1283.

14　R. B. Silva, H. Aldoradin-Cabeza, G. D. Eslick et al., "The Effect of Physical Exercise on Frail Older Persons: A Systematic Review," *Journal of Frailty Aging* 6, no. 2 (2017): 91–96.

15　R. D. Pollock, S. Carter, C. P. Velloso et al., "An Investigation into the Relationship between Age and Physiological Function in Highly Active Older Adults," *Journal of Physiology* 593, no. 3 (February 2015): 657–680; discussion, 680.

16　R. F. Gottesman, A. L. Schneider, M. Albert et al., "Midlife Hypertension and 20-Year Cognitive Change: The Atherosclerosis Risk in Communities Neurocognitive Study," *JAMA Neurology* 71, no. 10 (October 2014): 1218–1227.

17　K. A. Walker, M. C. Power, and R. F. Gottesman, "Defining the Relationship between Hypertension, Cognitive Decline, and Dementia: A Review," *Current Hypertension Reports* 19, no. 3 (March 2017): 24.

18 R. F. Gottesman, A. L. Schneider, Y. Zhou et al., "Association between Midlife Vascular Risk Factors and Estimated Brain Amyloid Deposition," *JAMA* 317, no. 14 (April 2017): 1443–1450.

19 K. Ding, T. Tarumi, D. C. Zhu et al., "Cardiorespiratory Fitness and White Matter Neuronal Fiber Integrity in Mild Cognitive Impairment," *Journal of Alzheimer's Disease* 61, no. 2 (2018): 729–739.

20 H. Arem, S. C. Moore, A. Patel et al., "Leisure Time Physical Activity and Mortality: A Detailed Pooled Analysis of the Dose-Response Relationship," *JAMA Internal Medicine* 175, no. 6 (June 2015): 959–967.

CHAPTER 5: THE POWER OF PURPOSE, LEARNING, AND DISCOVERY

1 C. Dufouil, E. Pereira, G. Chêne et al., "Older Age at Retirement Is Associated with Decreased Risk of Dementia," *European Journal of Epidemiology* 29, no. 5 (May 2014): 353–361.

2 R. Katzman, R. Terry, R. DeTeresa et al., "Clinical, Pathological, and Neurochemical Changes in Dementia: A Subgroup with Preserved Mental Status and Numerous Neocortical Plaques," *Annals of Neurology* 23 (1988): 138–144.

3 A. C. van Loenhoud, W. M. van der Flier, A. M. Wink et al., "Cognitive Reserve and Clinical Progression in Alzheimer Disease: A Paradoxical Relationship, *Neurology* 93, no. 4 (July 2019): e334–e346.

4 R. S. Wilson, L. Yu, M. Lamar et al., "Education and Cognitive Reserve in Old Age," *Neurology* 92, no. 10 (March 2019): e1041–e1050.

5 "Education May Not Protect against Dementia As Previously Thought" press release, February 6, 2019, American Academy of Neurology.

6 Kathleen Fifield, "College Education Doesn't Protect against Alzheimer's," AARP, February 6, 2019: www.aarp.org/health/dementia/info-2019/college -degree-dementia-prevention.html.

7 Laura Skufca, "2015 Survey on Brain Health," AARP Research, www.aarp.org /content/dam/aarp/research/surveys_statistics/health/2015/2015-brain -health.doi.10.26419%252Fres.00114.001.pdf.

8 T. H. Bak, J. J. Nissan, M. M. Allerhand et al., "Does Bilingualism Influence Cognitive Aging?" *Annals of Neurology* 75, no. 6 (June 2014): 959–963.

9 E. Bialystok, "Reshaping the Mind: The Benefits of Bilingualism," *Canadian Journal of Experimental Psychology* 65, no. 4 (December 2011): 229–235.

10 Jerri D. Edwards, Huiping Xu, Daniel O. Clark et al., "Speed of Processing Training Results in Lower Risk of Dementia," *Alzheimer's & Dementia* 3, no. 4 (November 2017): 603–611. Published online November 7, 2017.

11 L. G. Appelbaum, M. S. Cain, E. F. Darling et al., "Action Video Game Playing Is Associated with Improved Visual Sensitivity, But Not Alterations in Visual Sensory Memory," *Attention, Perception, and Psychophysics* 75, no. 6 (August 2013): 1161–1167.

12 J. A. Anguera, J. Boccanfuso, J. L. Rintoul et al., "Video Game Training Enhances Cognitive Control in Older Adults," *Nature* 501, no. 7465 (September 2013): 97–101. Also see https://neuroscape.ucsf.edu.

13 E. S. Kim, I. Kawachi, Y. Chen et al., "Association between Purpose in Life and Objective Measures of Physical Function in Older Adults," *JAMA Psychiatry* 74, no. 10 (October 2017): 1039–1045.

14 L. Yu, P. A. Boyle, R. S. Wilson et al., "Purpose in Life and Cerebral Infarcts in Community-Dwelling Older People," *Stroke* 46, no. 4 (April 2015): 1071–1076.

15 Global Council on Brain Health, "Brain Health and Mental Well-Being: GCBH Recommendations on Feeling Good and Functioning Well" (2018), www.GlobalCouncilOnBrainHealth.org.

16 Mihaly Csikszentmihalyi, *Flow: The Psychology of Optimal Experience* (New York: Harper & Row, 1990).

CHAPTER 6: THE NECESSITY OF SLEEP AND RELAXATION

1 For access to a library of resources and data about sleep, see the National Sleep Foundation's website: SleepFoundation.org.

2 Matthew Walker, *Why We Sleep: Unlocking the Power of Sleep and Dreams* (New York: Scribner, 2017).

3 See: https://aasm.org/resources/factsheets/sleepapnea.pdf.

4 S. Taheri, L. Lin, D. Austin, T. Young et al., "Short Sleep Duration Is Associated with Reduced Leptin, Elevated Ghrelin, and Increased Body Mass Index," *PLoS Medicine* 1, no. 3 (December 2004): e62.

5 J. G. Jenkins and K. M. Dallenbach, "Oblivescence During Sleep and Waking, *American Journal of Physiology* 35, no. 4 (October 1924): 605–12.

6 S. M. Purcell, D. S. Manoach, C. Demanuele et al., "Characterizing Sleep Spindles in 11,630 Individuals from the National Sleep Research Resource," *Nature Communications* 26, no. 8 (June 2017): 15930.

7   A. S. Lim, M. Kowgier, L. Yu et al., "Sleep Fragmentation and the Risk of Incident Alzheimer's Disease and Cognitive Decline in Older Persons," *Sleep* 36, no. 7 (July 2013): 1027–1032.

8   L. K. Barger, Shantha M.W. Rajaratnam, Christopher P. Cannon et al., "Short Sleep Duration, Obstructive Sleep Apnea, Shiftwork, and the Risk of Adverse Cardiovascular Events in Patients after an Acute Coronary Syndrome," *Journal of the American Heart Association* 6, no. 10 (October 2017): e006959.

9   C. W. Kim, Y. Chang, E. Sung, and S. Ryu, "Sleep Duration and Progression to Diabetes in People with Prediabetes Defined by HbA1c Concentration," *Diabetes Medicine* 34, no. 11 (November 2017): 1591–1598.

10   M. R. Irwin, M. Wang, D. Ribeiro et al., "Sleep Loss Activates Cellular Inflammatory Signaling," *Biological Psychiatry* 64, no. 6 (September 2008): 538–540.

11   K. A. Walker, R. C. Hoogeveen, A. R. Folsom et al., "Midlife Systemic Inflammatory Markers Are Associated with Late-Life Brain Volume: The ARIC Study," *Neurology* 89, no. 22 (November 2017): 2262–2270.

12   J. J. Iliff, M. Wang, Y. Liao et al., "A Paravascular Pathway Facilitates CSF Flow through the Brain Parenchyma and the Clearance of Interstitial Solutes, Including Amyloid β," in *Science Translational Medicine* 4, no. 147 (August 2012): 147ra111.

13   L. Xie, H. Kang, Q. Xu et al., "Sleep Drives Metabolite Clearance from the Adult Brain," *Science* 342, no. 6156 (October 2013): 373–377.

14   E. Shokri-Kojori, G. J. Wang, C. E. Wiers et al., "β-Amyloid Accumulation in the Human Brain after One Night of Sleep Deprivation," *Proceedings of the National Academy of Sciences USA* 115, no. 17 (April 2018): 4483–4488.

15   P. Li, Ing-Tsung Hsiao, Chia-Yih Liu et al., "Beta-Amyloid Deposition in Patients with Major Depressive Disorder with Differing Levels of Treatment Resistance: A Pilot Study," EJNMMI Res. 7, no. 1 (December 2017): 24; also see S. Perin, K. D. Harrington, Y. Y. et al., "Amyloid Burden and Incident Depressive Symptoms in Preclinical Alzheimer's Disease," *Journal of Affective Disorders* 229 (March 2018): 269–274.

16   Xie et al., "Sleep Drives Metabolite Clearance from the Adult Brain."

17   J. K. Holth, S. K. Fritschi, C. Wang et al., "The Sleep-Wake Cycle Regulates Brain Interstitial Fluid Tau in Mice and CSF Tau in Humans," *Science* 363, no. 6429 (2019): 880–884.

18   B. T. Kress, J. J. Iliff, M. Xia et al., "Impairment of Paravascular Clearance Path-

ways in the Aging Brain," *Annals of Neurology* 76, no. 6 (December 2014): 845–861.

19 A. P. Spira, L. P. Chen-Edinboro, M. N. Wu et al., "Impact of Sleep on the Risk of Cognitive Decline and Dementia," *Current Opinion Psychiatry* 27, no. 6 (November 2014): 478–483.

20 Jun Oh, Rana A. Eser, Alexander J. Ehrenberg et al., "Profound Degeneration of Wake-Promoting Neurons in Alzheimer's Disease," *Alzheimer's & Dementia* 15, no. 10 (2019): 1253–1263.

21 A. M. Chang, Daniel Aeschbach, Jeanne F. Duffy, and Charles A. Czeisler, "Evening Use of Light-Emitting eReaders Negatively Affects Sleep, Circadian Timing, and Next-Morning Alertness," *Proceedings of the National Academy of Sciences USA* 112, no. 4 (January 2015): 1232–1237.

22 Dr. Sanjay Gupta, CNN.com, 2017.

23 "Use of Yoga and Meditation Becoming More Popular in U.S.," press release, November 8, 2018, www.cdc.gov/nchs/pressroom/nchs_press_releases/2018/201811_Yoga_Meditation.htm.

24 Douglas C. Johnson, Nathaniel J. Thom, Elizabeth A. Stanley et al., "Modifying Resilience Mechanisms in At-Risk Individuals: A Controlled Study of Mindfulness Training in Marines Preparing for Deployment," *American Journal of Psychiatry* 171, no. 8 (August 2014): 844–853.

25 M. Goyal, S. Singh, E. M. Sibinga et al., "Meditation Programs for Psychological Stress and Well-Being: A Systematic Review and Meta-Analysis," *JAMA Internal Medicine* 174, no. 3 (March 2014): 357–368.

26 D. W. Orme-Johnson and V. A. Barnes, "Effects of the Transcendental Meditation Technique on Trait Anxiety: A Meta-Analysis of Randomized Controlled Trials," *Journal of Alternative and Complementary Medicine* 20, no. 5 (May 2014): 330–341.

27 S. W. Lazar, C. E. Kerr, R. H. Wasserman et al., "Meditation Experience Is Associated with Increased Cortical Thickness," *Neuroreport* 16, no. 17 (November 2005): 1893–1897.

28 Li Q, "Effect of Forest Bathing Trips on Human Immune Function," *Environmental Health and Preventive Medicine* 15, no. 1 (January 2010): 9–17.

29 M. M. Hansen, R. Jones, and K. Tocchini, "Shinrin-Yoku (Forest Bathing) and Nature Therapy: A State-of-the-Art Review," *International Journal of Environmental Research and Public Health* 14, no. 8 (July 2017): 851.

30 J. Barton and M. Rogerson, "The Importance of Greenspace for Mental Health," *The British Journal of Psychiatry* 14, no. 4 (November 2017): 79–81.

31 Kathleen Fifield, "New Report Finds Links between 'Mental Well-Being' and Brain Health," AARP, October 10, 2018, www.aarp.org/health/brain-health /info-2018/mental-well-being-connection-report.html.

32 Joel Wong and Joshua Brown, "How Gratitude Changes You and Your Brain," *Greater Good Magazine*, June 6, 2017, greatergood.berkeley.edu/article/item /how_gratitude_changes_you_and_your_brain.

33 Kirsten Weir, "Forgiveness Can Improve Mental and Physical Health," *Monitor on Psychology* 48, no. 1 (January 2017): 30.

34 D. G. Blanchflower and A. J. Oswald, "Is Well-Being U-Shaped over the Life Cycle?" *Social Science and Medicine* 66, no. 8 (April 2008): 1733–1749.

35 A. E. Reed and L. L. Carstensen, "The Theory behind the Age-Related Positivity Effect," *Frontiers in Psychology* 3 (September 2012): 339.

36 B. S. Diniz, M. A. Butters, S. M. Albert et al., "Late-Life Depression and Risk of Vascular Dementia and Alzheimer's Disease: Systematic Review and Meta-Analysis of Community-Based Cohort Studies," *The British Journal of Psychiatry* 202, no. 5 (May 2013): 329–335.

CHAPTER 7: FOOD FOR THOUGHT

1 S. Kahan and J. E. Manson, "Nutrition Counseling in Clinical Practice: How Clinicians Can Do Better," *JAMA* 318, no. 12 (September 2017): 1101–1102.

2 Kellie Casavale, "Promoting Nutrition Counseling as a Priority for Clinicians," Office of Disease Prevention and Health Promotion, November 29, 2017, www .health.gov.

3 S. B. Seidelmann, B. Claggett, S. Cheng et al., "Dietary Carbohydrate Intake and Mortality: A Prospective Cohort Study and Meta-Analysis," *Lancet* 3, no. 9 (September 2018): e419–e428.

4 Ramón Estruch, Emilio Ros, Jordi Salas-Salvadó et al., "Primary Prevention of Cardiovascular Disease with a Mediterranean Diet," *New England Journal of Medicine* 368, no. 14 (April 2013): 1279–1290.

5 Ramón Estruch, Emilio Ros, Jordi Salas-Salvadó et al., "Primary Prevention of Cardiovascular Disease with a Mediterranean Diet," *New England Journal of Medicine* 378, no. 25 (June 2018): e34.

6 M. C. Morris, C. C. Tangney, Y. Wang et al., "MIND Diet Associated with Re-

duced Incidence of Alzheimer's Disease," *Alzheimer's & Dementia* 11, no. 9 (September 2015): 1007–1014.

7 Martha Claire Morris, *Diet for the MIND: The Latest Science on What to Eat to Prevent Alzheimer's and Cognitive Decline* (New York: Little, Brown, 2017).

8 "AARP Releases Consumer Insights Survey on Nutrition and Brain Health," AARP, January 30, 2018, press.aarp.org/2018-1-30-AARP-Releases-Consumer-Insights-Survey-Nutrition-Brain-Health.

9 Richard Isaacson and Christopher Ochner, *The Alzheimer's Prevention and Treatment Diet* (Garden City Park, NY: Square One, 2016).

10 R. S. Isaacson, C. A. Ganzer, H. Hristov et al., "The Clinical Practice of Risk Reduction for Alzheimer's Disease: A Precision Medicine Approach," *Alzheimer's & Dementia* 14, no. 12 (December 2018): 1663–1673.

11 Richard Isaacson, Hollie Hristov, Nabeel Saif et al., "Individualized Clinical Management of Patients at Risk for Alzheimer's Dementia," *Alzheimer's & Dementia*, October 30, 2019, www.alzheimersanddementia.com/articleS1552-5260(19)35368-3/fulltext.

12 For more about Dean Ornish's research and works, go to www.ornish.com.

13 S. Kahan and J. E. Manson, "Nutrition Counseling in Clinical Practice: How Clinicians Can Do Better," *JAMA* 318, no. 12 (September 2017): 1101–1102.

14 C. D. Fryar, J. P. Hughes, K. A. Herrick, and N. Ahluwalia, "Fast Food Consumption among Adults in the United States, 2013–2016," National Center for Health Statistics data brief 322, 2018.

15 "AARP Releases Consumer Insights Survey on Nutrition and Brain Health," AARP, January 30, 2018, press.aarp.org/2018-1-30-AARP-Releases-Consumer-Insights-Survey-Nutrition-Brain-Health.

16 U.S. Department of Agriculture, Economic Research Service, "Food Availability and Consumption," accessed October 28, 2019, www.ers.usda.gov/data-products/ag-and-food-statistics-charting-the-essentials/food-availability-and-consumption/.

17 Jotham Suez, Tal Korem, David Zeevi et al., "Artificial Sweeteners Induce Glucose Intolerance by Altering the Gut Microbiota," *Nature* 514 (Ocober 2014): 181–186.

18 M. T. Wittbrodt and M. Millard-Stafford, "Dehydration Impairs Cognitive Performance: A Meta-Analysis," *Medicine and Science in Sports and Exercise* 50, no. 11 (November 2018): 2360–2368.

19 S. C. Larsson and N. Orsini, "Coffee Consumption and Risk of Dementia and Alzheimer's Disease: A Dose-Response Meta-Analysis of Prospective Studies," *Nutrients* 10, no. 10 (October 2018): 1501.

20 Bridget F. Grant, S. Patricia Chou, Tulshi D. Saha et al., "Prevalence of 12-Month Alcohol Use, High-Risk Drinking, and DSM-IV Alcohol Use Disorder in the United States, 2001–2002 to 2012–2013: Results from the National Epidemiologic Survey on Alcohol and Related Conditions," *JAMA Psychiatry* 74, no. 9 (September 2017): 911–923.

21 J. E. Manson, N. R. Cook, I. M. Lee et al., "Marine n-3 Fatty Acids and Prevention of Cardiovascular Disease and Cancer," *New England Journal of Medicine* 380, no. 1 (January 2019): 23–32.

22 J. I. Fenton, N. G. Hord, S. Ghosh, and E. A. Gurzell, "Immunomodulation by Dietary Long Chain Omega-3 Fatty Acids and the Potential for Adverse Health Outcomes," *Prostaglandins, Leukotrienes and Essential Fatty Acids* 89, no. 6 (November–December 2013): 379–390.

23 Dean Sherzai and Ayesha Sherzai, *The Alzheimer's Solution: A Breakthrough Program to Prevent and Reverse the Symptoms of Cognitive Decline at Every Age* (San Francisco: HarperOne, 2017).

24 Joe Sugarman, "Are There Any Proven Benefits to Fasting?" *Johns Hopkins Health Review* 3, no. 1 (Spring/Summer 2016), 9–10.

25 M. P. Mattson, V. D. Longo, and M. Harvie, "Impact of Intermittent Fasting on Health and Disease Processes," *Ageing Research Reviews* 39 (October 2017): 46–58.

26 M. P. Mattson, K. Moehl, N. Ghena et al., "Intermittent Metabolic Switching, Neuroplasticity and Brain Health," *Nature Reviews Neuroscience* 19, no. 2 (February 2018): 6–80.

27 Mayo Clinic Staff, "Dietary Fiber: Essential for a Healthy Diet," accessed October 28, 2019, www.mayoclinic.org.

28 G. W. Small, P. Siddarth, Z. Li et al., "Memory and Brain Amyloid and Tau Effects of a Bioavailable Form of Curcumin in Non-Demented Adults: A Double-Blind, Placebo-Controlled 18-Month Trial," *American Journal of Geriatric Psychiatry* 26, no. 3 (March 2018): 266–277.

29 B. Lebwohl, Y. Cao, G. Zong et al., "Long Term Gluten Consumption in Adults without Celiac Disease and Risk of Coronary Heart Disease: Prospective Cohort Study," *British Medical Journal* 2, no. 357 (2017): j1892.

CHAPTER 8: CONNECTION FOR PROTECTION

1 J. Holt-Lunstad, T. F. Robles, and D. A. Sbarra, "Advancing Social Connection as a Public Health Priority in the United States," *American Journal of Psychology* 72, no. 6 (September 2017): 517–530.

2 H. Liu, Z. Zhang, S. W. Choi, and K. M. Langa, "Marital Status and Dementia: Evidence from the Health and Retirement Study," *Journals of Gerontology, Series B: Psychological Sciences and Social Sciences* (June 2019): gbz087.

3 Sharon M. Lee and Barry Edmonston, "Living Alone Among Older Adults in Canada and the U.S." *Healthcare* (Basel) 7, no. 2 (June 2019): 68. Also see: Dhruv Khullar, "How Social Isolation Is Killing Us," *The New York Times* The Upshot section, December 22, 2016.

4 "AARP Survey Reveals Being Social Promotes Brain Health," AARP press room, March 28, 2017, press.aarp.org/2017-03-28-AARP-Survey-Reveals-Being-Social-Promotes-Brain-Health.

5 N. K. Valtorta, M. Kanaan, S. Gilbody et al., "Loneliness and Social Isolation as Risk Factors for Coronary Heart Disease and Stroke: Systematic Review and Meta-Analysis of Longitudinal Observational Studies," *Heart* 102, no. 13 (July 2016): 1009–1016.

6 J. Holt-Lunstad, T. B. Smith, M. Baker et al., "Loneliness and Social Isolation as Risk Factors for Mortality: A Meta-Analytic Review," *Perspectives on Psychological Science* 10, no. 2 (March 2015): 227–237.

7 Kassandra I. Alcaraz, Katherine S. Eddens, Jennifer L. Blase et al., "Social Isolation and Mortality in U.S. Black and White Men and Women," *American Journal of Epidemiology* 188, no. 1 (November 2018): 102–109.

8 Michelle C. Carlson, Kirk I. Erickson, Arthur F. Kramer et al., "Evidence for Neurocognitive Plasticity in At-Risk Older Adults: The Experience Corps Program," *Journal of Gerontology: Medical Sciences* 64, no. 12 (December 2009): 1275–1282.

9 I. M. McDonough, S. Haber, G. N. Bischof, and D. C. Park, "The Synapse Project: Engagement in Mentally Challenging Activities Enhances Neural Efficiency," *Restorative Neurology and Neuroscience* 33, no. 6 (2015): 865–882.

10 D. A. Bennett, J. A. Schneider, A. S. Buchman et al., "Overview and Findings from the Rush Memory and Aging Project," *Current Alzheimer Research* 9, no. 6 (July 2012): 646–663.

11 Sanjay Gupta, "Just Say Hello: The Powerful New Way to Combat Loneliness," www.Oprah.com, February 18, 2014; http://www.oprah.com/health/just-say -hello-fight-loneliness/all#ixzz6BsFWtzlq.

12 Cigna U.S. Loneliness Index, 2018. www.multivu.com/players/English/829 4451-cigna-us-loneliness-survey/docs/IndexReport_1524069371598-1735 25450.pdf.

13 N. I. Eisenberger, M. D. Lieberman, and K. D. Williams, "Does Rejection Hurt? An FMRI Study of Social Exclusion," *Science* 302, no. 5643 (October 2003): 290–292.

14 See AdultDevelopmentStudy.org.

15 See Waldinger's 2015 TED talk: www.ted.com/speakers/robert_waldinger.

16 Ibid.

17 O. P. Almeida, B. B. Yeap, H. Alfonso et al., "Older Men Who Use Computers Have Lower Risk of Dementia," *PLoS One* 7, no. 8 (August 2012): e44239.

18 Janelle Wohltmann of the University of Arizona has been conducting this on- going study. She presented these findings at the International Neuropsycholog- ical Society Annual Meeting in 2013. www.tucsonsentinel.com/local/report /022013_facebook_for_seniors/ua-study-facebook-use-gives-seniors-cogni tive-boost/.

CHAPTER 10: DIAGNOSING AND TREATING AN AILING BRAIN

1 "Self-Reported Increased Confusion or Memory Loss and Associated Func- tional Difficulties Among Adults Aged ≥60 Year—21 States, 2011," *Morbid- ity and Mortality Weekly Report*, May 10, 2013, www.cdc.gov/mmwr/preview /mmwrhtml/mm6218a1.htm.

2 Sandee LaMotte and Stephanie Smith, "Sandy's Story: Fighting Alzheimer's," CNN Health, www.cnn.com/2015/10/12/health/Alzheimers=sandys-story?.

3 See www.alzdiscovery.org.

4 See www.alz.org.

5 Teresa Carr, "Too Many Meds? America's Love Affair With Prescription Med- ication," *Consumer Reports*, August 3, 2017. Numbers are based on a survey of nearly 2,000 Americans.

6 C. A. C. Coupland, T. Hill, T. Dening et al., "Anticholinergic Drug Exposure and the Risk of Dementia: A Nested Case-Control Study," *JAMA Internal Medicine* 179, no. 8 (June 2019): 1084–1093.

7 Somayeh Meysami, Cyrus A. Raji, David A. Merrill et al., "MRI Volumetric Quantification in Persons with a History of Traumatic Brain Injury and Cognitive Impairment," *Journal of Alzheimer's Disease* (August 2019): 1–8.

8 Elham Mahmoudi, Tanima Basu, Kenneth Langa et al., "Can Hearing Aids Delay Time to Diagnosis of Dementia, Depression, or Falls in Older Adults," *Journal of the American Geriatric Society* 67, no. 11 (November 2019): 2362–2369.

9 R. Brookmeyer and N. Abdalla, "Estimation of Lifetime Risks of Alzheimer's Disease Dementia Using Biomarkers for Preclinical Disease," *Alzheimer's & Dementia* 14, no. 8 (August 2018): 981–988.

CHAPTER 11: NAVIGATING THE PATH FORWARD FINANCIALLY AND EMOTIONALLY, WITH A SPECIAL NOTE TO CAREGIVERS

1 See longtermcare.acl.gov/costs-how-to-pay/costs-of-care.html.

2 For facts and figures on people living with dementia and their caregivers, see the Alzheimer's Association website: www.alz.org/media/documents/alzheimers-facts-and-figures-2019-r.pdf.

3 alz.org; mybrain.alz.org/alzheimers-facts.asp?_ga=2.131831943.961943911.1572215697-1067122304.1571678924.

4 Rainville et al., Family Caregiving and Out-of-Pocket Costs: 2016 Report. Washington, DC: AARP Research, Nov. 2016. doi.org/10.26419/res.00138.001.

5 Ensocare, "The High Cost of Forgoing Advance Directives," June 15, 2017, www.ensocare.com/knowledge-center/the-high-cost-of-forgoing-advance-directives.

6 Ibid.

7 Maria C. Norton, Ken R. Smith, Truls Ostbye et al., "Greater Risk of Dementia When Spouse Has Dementia? The Cache County Study," *Journal of the American Geriatric Society* 58, no. 5 (2010): 895–900.

Index

About the Author

Sanjay Gupta fell in love with the brain as a young boy in middle school. He later went on to spend four years earning a medical degree and then seven years completing residency training so he could become a neurosurgeon—a practice he has been enjoying for the last twenty-some years. The brain is his first and truest love.

Dr. Gupta is a three-time *New York Times* bestselling author and serves as the chief medical correspondent for CNN. Since 2001, Gupta has covered the biggest health headlines of our time—often telling the harrowing and touching stories of brave first responders, and reporting from the front lines of nearly every conflict, natural disaster, and disease outbreak anywhere in the world. He has hosted several long-form documentaries based on deep investigations, including his *Weed* series and *One Nation Under Stress* for HBO. For his work, he has achieved multiple Emmy and Peabody awards, as well as the DuPont award—the broadcast equivalent of the Pulitzer. To write his nonfiction books, *Chasing Life and Cheating Death*, Gupta gathered stories by traveling the world looking at long-lived cultures and societies pushing back the boundaries of death.

Gupta is widely regarded as one of the most trusted reporters in the media. In addition to his accolades in journalism, Gupta is the recipient of several honorary degrees and has been recognized with many humanitarian awards for his care of people injured in wars and natural disasters. *Forbes* magazine named him one of the ten most influential celebrities.

In 2019, Gupta was elected to the National Academy of Medicine, one of the highest honors in the medical field.

Gupta lives in Atlanta where he is also an associate professor of neurosurgery at Emory University Hospital and associate chief of neurosurgery at Grady Memorial Hospital. He serves as a diplomate of the American Board of Neurosurgery. Sanjay is married to Rebecca, who, after reading this, reminded him that she was in fact his truest love. He wisely conceded this point. They have three teen and preteen daughters, who find it hilarious their father is writing a book about memory. As they put it, the Gupta Girls believe their father "literally can't remember anything."